National Academy Press • 2101 Constitution Avenue, N.W. • Washington, D.C. 20418

NOTICE: The National Academy of Engineering was established in 1964, under the charter of the National Academy of Sciences, as a parallel organization of outstanding engineers. It is autonomous in its administration and in the selection of its members, sharing with the National Academy of Sciences the responsibility for advising the federal government. The National Academy of Engineering also sponsors engineering programs aimed at meeting national needs, encourages education and research, and recognizes the superior achievement of engineers. Dr. Robert M. White is president of the National Academy of Engineering.

This publication has been reviewed by a group other than the authors according to a National Academy of Engineering report review process. Inclusion of signed work in this publication signifies that it is judged a competent and useful contribution worthy of public consideration, but it does not imply endorsement of conclusions or recommendations by the National Academy of Engineering. The interpretations and conclusions expressed in this volume are those of the authors and are not presented as the views of the council, officers, or staff of the Academy.

Partial funding for this effort was provided by the Andrew W. Mellon Foundation and the National Academy of Engineering Technology Agenda Program.

Library of Congress Cataloging-in-Publication Data

Energy : production, consumption, and consequence / John L. Helm, editor.
    p. cm.
    "National Academy of Engineering."
    Companion volume to: Technology and environment.
    Based on a symposium entitled An energy agenda for the 1990s, held at the Arnold and Mabel Beckman Center of the National Academy of Sciences and Engineering, Irvine, Calif., sponsored by the Program Office of the National Academy of Engineering.
    Includes bibliographical references.
    ISBN 0-309-04077-9
    1. Power resources—Congresses. I. Helm, John L. II. National Academy of Engineering. Program Office. III. Technology and environment.
TJ163.15.E5384   1989
333.79—dc20
                                                89-13364
                                                CIP

Cover: Toni Simon, *Fusion*, oil on canvas. Courtesy of the artist.

Copyright © 1990 by the National Academy of Sciences

Printed in the United States of America

ENERGY
PRODUCTION, CONSUMPTION,
AND CONSEQUENCES

John L. Helm
Editor

National Academy of Engineering

NATIONAL ACADEMY PRESS
Washington, D.C.   1990

Symposium Advisory Committee

WILLIAM F. KIESCHNICK, *Chairman*, Director and Chief Executive Officer (retired), Atlantic Richfield Company
RICHARD E. BALZHISER, President, Electric Power Research Institute
A. PHILIP BRAY, Vice Chairman and Managing Director, Renewable Resource Systems, Inc.
HARVEY BROOKS, Professor of Technology and Public Policy, Emeritus, Kennedy School of Government
THEODORE A. BURTIS, Director, Sun Company, Inc.
FLOYD L. CULLER, President Emeritus, Electric Power Research Institute
W. KENNETH DAVIS, Former Deputy Secretary, Department of Energy
FREDERICK J. ELLERT, President, Ellert Consulting Group, Inc.
JAMES L. EVERETT III, Chairman and Chief Executive Officer (retired), Philadelphia Electric Company
ROBERT A. FROSCH, Vice President, General Motors Research Laboratories, General Motors Corporation
JOHN H. GIBBONS, Director, Office of Technology Assessment
WILLIAM R. GOULD, Chairman Emeritus, Southern California Edison Company
MICHEL T. HALBOUTY, Chairman and Chief Executive Officer, Michel T. Halbouty Energy Company
FRED L. HARTLEY, Chairman Emeritus, Unocal Corporation
WILLIS M. HAWKINS, Senior Adviser, Lockheed Corporation
THOMAS H. LEE, Professor, Massachusetts Institute of Technology
WILLIAM S. LEE, Chairman and Chief Executive Officer, Duke Power Company
HENRY R. LINDEN, Executive Adviser, Gas Research Institute
PLATO MALOZEMOFF, Chairman Emeritus, Newmont Mining Corporation
WALTER J. McCARTHY, Jr., Chairman and Chief Executive Officer, Detroit Edison Company
RICHARD M. MORROW, Chairman and Chief Executive Officer, Amoco Corporation
WILLIAM A. NIERENBERG, Director Emeritus, Scripps Institute of Oceanography
THOMAS H. PIGFORD, Professor of Nuclear Engineering, University of California, Berkeley
ERIC H. REICHL, President (retired), Conoco Coal Development Company

DONALD G. RUSSELL, Chairman and Chief Executive Officer, Sonat Exploration Company
GLENN A. SCHURMAN, Retired Vice President, Chevron Corporation
CHAUNCEY STARR, President Emeritus, Electric Power Research Institute
JOHN E. SWEARINGEN, Retired Chairman, Standard Oil Company (Indiana)
JOHN A. TILLINGHAST, President, TILTEC
RICHARD F. TUCKER, President, Mobil Oil Corporation
ALVIN M. WEINBERG, Distinguished Fellow, Institute for Energy Analysis, Oak Ridge Associated Universities
DAVID C. WHITE, Ford Professor of Engineering, Director Energy Laboratory, Massachusetts Institute of Technology
HERBERT H. WOODSON, Dean of Engineering, University of Texas
ALDEN P. YATES (deceased), President, Bechtel Group, Inc.

Preface

The global supply of fuels and other sources of energy is more than adequate to meet present needs. The deliverable supply, however, is constrained by the interplay of public policy, technology, environmental concerns, and economic forces. Over the past two decades, the focus of effort has been on managing the deliverable supply of energy in the face of sometimes opposing global and domestic forces.

In this volume a group of leading authorities on energy and related environmental issues explores some of these emerging and evolving forces. The organizers of the symposium on which this volume is based have opted to examine these issues in a framework different from the traditional one, usually structured by supply sector. Instead, analysis of the energy system has been approached from the perspective of demand and supply interactions, environmental effects, and evolving vulnerabilities and opportunities. Implications for energy strategy have been distilled from these perspectives. This approach prevented the synoptic coverage of all the possible supply sectors; topics such as coal-based technologies and solar energy systems have not been addressed in depth.

Several major themes emerge from this volume. Of significance are the views of how supply and demand factors interact and the influence of technology on them. The shaping of demand by the growing role of electricity is discussed. The geopolitics of development, transportation, marketing, use of fossil fuels, and their regional and global environmental effects is the framework for examining the vulnerabilities of energy price volatility, the uncertain future role of methane, and possible consequences of continued disarray in the nuclear industry.

The strongest theme in this volume is the growing importance of environmental concerns in planning the global energy system. Paradoxically, the pursuit of abundant supplies of energy, so vital to the economic well-being and to the general welfare of humankind, creates the dilemma that present modes of energy use also threaten serious environmental deterioration. Our knowledge of the causes and consequences of various environmental phenomena varies widely. However, it is becoming clear that the production, distribution, and consumption of energy by industrial societies generate unwelcome planetary environmental loadings. This volume explores this relationship. A companion book of the National Academy of Engineering, *Technology and Environment* (J. Ausubel and H. Sladovich, eds.) explores more broadly how technology can be used to manage humanity's ever more intimate relationship with the environment.

This book is based on the symposium "An Energy Agenda for the 1990s," the first of an inaugural series of symposia to celebrate the opening of the Arnold and Mabel Beckman Center of the National Academies of Sciences and Engineering in Irvine, California. The symposium and planning effort were guided by William F. Kieschnick, who chaired the symposium advisory committee and a steering committee consisting of Richard E. Balzhiser, W. Kenneth Davis, John H. Gibbons, Thomas H. Lee, Henry R. Linden, Glenn A. Schurman, and John A. Tillinghast, to whom we are indebted. The committee was assisted by NAE fellow John L. Helm, who served as editor of the symposium papers that have come to make up this volume. This project was carried out under the auspices of the Program Office of NAE, directed by Jesse H. Ausubel and more recently by Bruce R. Guile. We are also indebted to Jay Ball for assistance in organizing the symposium and related activities, to Samuel R. Rod for assistance in the review process, and to H. Dale Langford and Bette R. Janson for their work in preparing the manuscript for publication.

> ROBERT M. WHITE
> President
> National Academy of Engineering

Contents

Energy Planning in a Dynamic World: Overview and Perspective 1
William F. Kieschnick and John L. Helm

1. SUPPLY, DEMAND, AND REAPPRAISAL

Energy in Retrospect: Is the Past Prologue? 21
Alvin M. Weinberg

Energy Efficiency: Its Potential and Limits to the Year 2000 35
John H. Gibbons and Peter D. Blair

Implications of Continuing Electrification 52
Chauncey Starr

2. ENVIRONMENTAL ISSUES

Global Environmental Forces 75
Thomas C. Schelling

Regional Environmental Forces: A Methodology for
Assessment and Prediction 85
Thomas E. Graedel

The Automobile and the Atmosphere 111
John W. Shiller

3. EVOLVING VULNERABILITIES AND OPPORTUNITIES

Managing Volatility in the Oil Industry 145
    John F. Bookout

The Uncertain Future Role of Natural Gas 165
    William T. McCormick, Jr.

European Natural Gas Supplies and Markets 173
    Henrick Ager-Hanssen

Future Consequences of Nuclear Nonpolicy 184
    Richard E. Balzhiser

4. IMPLICATIONS FOR STRATEGY

Energy, Environment, and Development 205
    William D. Ruckelshaus

What to Do About CO_2 213
    John L. Helm and Stephen H. Schneider

Achieving Continuing Electrification 238
    Wallace B. Behnke, Jr.

Regional Approaches to Transboundary Air Pollution 246
    Peter H. Sand

Efficiency, Machiavelli, and Buddha 265
    Robert Malpas

Contributors 279

Index 285

Energy Planning in a Dynamic World: Overview and Perspective

WILLIAM F. KIESCHNICK AND JOHN L. HELM

> Our preference for muddling through has had its day, and it will not be good enough for the third century of the Constitution.
>
> *William D. Carey*, former executive officer, American Association for the Advancement of Science

In the 16 years since the oil shock of 1973, many individuals, institutions, and government bodies have directed much effort toward revitalizing and fortifying the energy system in the United States. Although much progress has been achieved, the formulation of a coherent long-range national energy policy has not yet been accomplished. Nonetheless, actively planning for the country's future energy system seems to have fallen in importance relative to other issues on the national agenda. The authors in this volume make clear that the time is at hand for a reappraisal of the way in which we plan for and use our national energy system.

The agenda for energy institutions has always been to provide energy in the forms required to support economic growth, social progress, and ever-changing societal needs and values. In effect, there has been a social contract between provider, consumer, and government. Consumers, needing energy for existence and enterprise, set specifications by their needs and value systems and seek to pay the lowest possible price for delivery. Providers continually assess the consumers' needs and values, invest in projects and programs to deliver energy, and receive consumer payments to underwrite their expenses and capital. Government serves as the referee and auditor of the process and intervenes to set ground rules, provide incentives, and mitigate emergencies.

This agenda has been expressed in a variety of settings over time. It is particularly interesting and useful to compare the energy context of the 1970s with that emerging for the 1990s. The key features of the 1970s included the following:

- Rapidly rising prices.
- Widely held beliefs that both prices and consumption would continue to rise without moderation (no significant price elasticity).
- Perceived episodic shortages leading to real episodic shortages. Rising consumption exacerbated by panic purchases. Government interventions to allocate scarce supplies.
- Major emphasis on supply management; all viable energy sources should be mobilized and alternative energy sources should be developed for the future.
- Government mandates to achieve energy demand reduction by efficiency standards for automobiles, appliances, commercial buildings, and residences.
- Strong market and government incentives for investment in long-term energy supply.
- Tacit assumptions that environmental problems would remain local in scope and character and that their solution would be straightforward.
- Acute concern over economic and strategic vulnerabilities due to high petroleum prices and dependence on foreign supplies.

The emerging market conditions of the 1990s differ significantly:

- Volatile prices oscillating around a gradually changing mean price, probably rising in real terms.
- Increased recognition of price elasticities in the free markets, and the realization that these elasticities are constructive and tend to moderate emergencies.
- Shifts from supply shortages to supply excess and rising consumption.
- More sophisticated understanding of demand management and how demand can be mitigated by efficiency measures and how price influences supply enlargement and diversification by alternative fuels.
- Government deregulation of supply sectors such as the electric power distribution grid and the natural gas pipelines.
- Disincentives for long-range energy investments, such as energy commodity price volatility and short time horizons for competing capital investment opportunities.

OVERVIEW AND PERSPECTIVE 3

- Mounting concerns about energy system impacts on the environment at global and regional scales as well as at the local level.
- Growing concern that the U.S. energy system will eventually be vulnerable again to price shocks, given increasing consumption and diminishing development of new energy sources because of currently lower energy prices.

The differences between the conditions of the 1970s and those of the 1990s are substantial. In the 1970s the perceived problems of the U.S. supply shortage and dollar drain were clear; high prices induced large investments in energy research and development and production; the environmental performance of energy systems was not confining. In the 1990s the potential for price and environmental shocks is largely speculative with respect to both timing and impact, yet anticipatory investment appears to be required if adverse contingencies are to be effectively mitigated. It is this changed context for energy planning that makes a fundamental reappraisal so important and thus presents an expanded set of challenges to our national energy future.

- How can energy managers, private and public, take advantage of our improved understanding about the interdependence of energy supply and demand with innovation and efficiency technologies?
- The environmental consequences of energy production, distribution, and use have grown to a planetary scale. What are the implications of this "environmental closure" for domestic and international energy planning and policy?
- What technological opportunities exist to respond to environmental closure, improve efficiency, increase supplies, and finesse uncertainty? What, if any, institutional measures will be required to promote the selection and deployment of such technologies?
- What private and public strategies are most likely to be effective in the context of a high degree of uncertainty about energy demand growth; the workings of our natural world and the ecosystems that support humankind; and the future role of technology, markets, and human ingenuity?

These questions are addressed by the chapters in this volume and are discussed in overview below.

SUPPLY, DEMAND, AND REAPPRAISAL

During the 1970s, most energy planning was premised on the prevailing assumption that energy demand was price-inelastic and on the expectation of high, continually escalating energy prices. As Alvin Weinberg explains, most studies before the 1980s significantly underestimated the long-term price elasticity of energy demand, the growing importance of the environment, the consequences of electrification, and the potential role of efficiency technologies. This happened partly because forecasters possessed little experience with surges in energy prices before the 1970s. What policymakers and energy analysts learned in the 1980s was how subtle and complex is the relationship between energy supply and demand and how sensitive it is to factors external to the energy system.

Despite growing knowledge of the relationship between energy supply and demand and the external feedbacks affecting energy production and consumption, it remains difficult to construct useful models of this system. At least two reasons for this difficulty emerge from the contributions in this volume.

First, changes can occur very rapidly. Perhaps the most striking example of this fact in the energy sector is given by John Bookout, who shows that the volatile price behavior of oil is typical of other commodities such as grains or metals. Weinberg shows rapid changes in the ratio of energy use to gross domestic product. Thomas Schelling observes that rapid change and adaptation are an inherent characteristic of humans and human systems, for example, people readily adapt to seasonal variations, different regional climates, and more. Hence, the role of rates of change is complex; although rapid change may promote adaptation, it has repeatedly undermined forecasting and planning. This consideration is especially true for the case of oil price volatility, because oil price serves as the yardstick by which energy planners evaluate other energy investments.

Second, the energy system has increasingly complex interconnections with other national and global structures such as the economy and the physical environment. These interconnections can sometimes lead to surprises and heightened vulnerability.

The rapid energy price rises of the 1970s in the United States contributed to general inflation and substantial capital transfers from oil-consuming to oil-producing nations. The effects on oil-importing nations were somewhat reversed when demand proved to be responsive to price and the oil crisis of the 1970s created enough change in the behavior of consumers to substantially reduce the rate of growth of energy demand. During the mid-1980s, U.S. consumers saw gasoline prices fall and stay low over long periods. Then, producer nations' economies suffered as a result of diminished world petroleum consumption. Both players—oil importers and exporters—experienced unexpected economic interactions.

Thomas Graedel shows clearly how complex interactions arise from the growing interdependence of the energy system and the environment. Even though energy generation has always produced undesirable by-products, until recently fuel use and the consequent pollution were thought to be relatively modest compared with the environment's ability to absorb the emissions. We are now recognizing such effects as local ozone buildup (Shiller, this volume), regional acid rain (Graedel, this volume) and the global greenhouse (Schelling, and Helm and Schneider, this volume), as symptoms of escalating energy consumption. Advances in understanding of the energy-environment connection present new and difficult trade-offs between the social value of energy and the social value of environmental quality.

Harnessing all sources of energy affects the physical and economic environment. Every energy option challenges policymakers with a set of social trade-offs, yet the political process for achieving, or even clearly understanding, social trade-offs is difficult. Any fossil fuel necessarily generates oxides of carbon and nitrogen when burned. Any long-distance transportation system for oil or gas has risks of occasional leaks and spills. Hydropower involves intrusions on natural water flow systems. Light-water atomic fission reactors have the potential for self-destructive transients, and all fission reactors produce spent fuel, the disposal of which requires centuries-long confinement. Solar power requires large amounts of materials whose manufacture involves mining and use of energy. Considering the diverse and pervasive nature of these effects of energy generation and use, energy planning involves getting the most effective and desirable energy for the least adverse impact.

The trade-off process is difficult at all levels of the energy system. As John Gibbons and Peter Blair show, individuals weigh the energy efficiency of consumer goods against price and competing features on almost a daily basis. John Shiller explains that as the need to improve the environmental performance of consumer goods, such as motor vehicles, continues to grow, so too will the complexity of the trade-offs required at the time of purchase.

The difficulty of the trade-off process can also be observed at the institutional level in Wallace Behnke's discussion of future electricity supply issues; grid reliability, supply dependability, open access, and deregulation, among other considerations, are placing complex and often competing demands on the electric power industry. Richard Balzhiser calls for the institutionalization of the trade-off process to establish the "risk/reward symmetry" that he asserts is critical to the repair of the nuclear industry. The challenge and the necessity of achieving trade-offs at the international level are presented by Peter Sand and William Ruckelshaus. Sand presents the principles on which transboundary air pollution has been managed over large regions, such as Europe and North America; while fairly straightforward in principle, the trade-off process among various nations can be intricate in practice. Ruckelshaus emphasizes that multilateral action is needed on a global scale to prevent the "pollute and cure" approach and to promote the "sustainable development" approach; even more difficult local versus global trade-offs will be required.

As populations grow and as the search for raw energy widens, interactions will grow and social trade-offs will become more difficult. This process has been especially visible in the United States in the public debate over control of urban ozone in a few cities, the production and transportation of oil in Alaska, and the exploration for oil and gas offshore in the United States. The social trade-offs to be made between constructive energy use and reasonable energy prices, on the one hand, and environmental quality control, on the other hand, will remain enormously difficult and challenging.

Another challenge to energy managers comes from the short-term planning bias that is inherent in energy prices. Robert Malpas discusses prices as "the criteria by which investment decisions are judged" and asserts that they do not adequately reflect the future costs and long-term consequences of current investments in terms of their potential, indeed inevitable, effect on the ecosystem and resource supplies. He attributes this to a mismatch of long time scales on the supply side to short time scales on the demand side. The long time scales inherent in managing energy supply arise from, for example, long planning and construction lead times and project lifetimes of 15 to 20 years or more. In contrast, the time scales on which decisions affecting energy demand are made daily by millions of individuals and organizations are based on short usage periods, typically 3 to 6 years. Further, energy institutions, policymakers, providers, and consumers understandably have different insights and beliefs about such complex phenomena as global trade and the greenhouse effect, which even experts understand with only limited precision. This leads to a natural bias toward short time frames and incremental change. The net result is that nonpollution and efficiency are undervalued and energy supply investments are made in long-lived, less efficient, or less environmentally benign technology. Malpas concludes that "The challenge facing demand-driven energy policy options is how to influence short-term decisions to take into account long-term opportunity and potential penalties."

Thus, planning for the future requires developing an adequate perception of energy futures and technological opportunities. Clearly energy futures frequently turn up surprises, but sometimes the surprise is due to the accumulated effects of long-term processes, which lead to an unstable system vulnerable to shocks, such as the embargo imposed in 1973 by the Organization of Petroleum Exporting Countries (OPEC). At other times, unexpected events trigger actions and reactions that lead to surprises such as the price elasticity of demand and the temporary gasoline deliverability shortages of the 1970s. Surprises also result because the model or method used to look to the future is only approximate and may not account for all that becomes important. OPEC-watchers have a poor record in predicting the outcome and impact of the sociological and political behavior of the off-and-on-again cartel. Scientists investigating the greenhouse phenomenon are concerned about the approximate nature of their computer models and about unpredictable instabilities—surprises—developing in the climate system, such as more rapid changes in global climate than are currently anticipated.

The chapters in this volume suggest that the most pervasive class of surprise that has frustrated forecasting has been provided by technology. Part of the challenge of accounting for the role of technology in energy planning arises because of its diverse embodiments. In addition to the large base of existing technologies that have been or can be selected for use and can be continuously improved, new and completely unpredictable technologies are continuously produced by the ingenuity of the innovators and inventors. The possibilities for unforeseen breakthroughs include renewable energy sources such as solar, fusion systems, fuel cells, and things yet to be imagined. How can a forecaster or planner account for technological serendipity in the energy system? To do so explicitly is confounded by several factors: the logic of resource scarcity is simple and direct, whereas the logic of resource expansion by human resourcefulness is complex and indirect; there are always short-term negative effects due to increased pressure on the current system, whereas the benefits that may result from the constructive adversity of this pressure come only later; and innovation often leads to changing the order of things and often includes individual, institutional, and even societal dislocations. Planning and reappraisal must become more sensitive to these considerations.

THE ENVIRONMENTAL DIMENSION

Almost all human activity leads to the discharge of by-products; and the production and use of energy, which itself is basic to all human activity, is no exception. With the explosive growth in energy consumption during the past century, there has necessarily been correspondingly rapid growth in the release of combustion products. The earth's atmosphere now contains about 20 percent more carbon dioxide than in 1900 and 25 percent more than before the industrial revolution 200 years ago. The scope of interactions between the energy system and the environment is unprecedented. Because of the long time frames involved and the growing potential for surprise, the reappraisal of energy systems worldwide has become an urgent priority.

In 1900 the world consumption of energy was about 1 billion gigawatts, and now is approaching 10 billion gigawatts. This tenfold increase in one century is the product of a threefold increase in world population and a roughly threefold increase in average per capita use. As Chauncey Starr argues, the increase in per capita energy use is fundamentally linked to the growth of the world economy. Starr predicts that as technological advance and evolving social organization combine to improve the standard of living in industrialized societies and to extend improvement to more societies, per capita energy demand will continue to grow. Because of lead times in deploying new energy technology, Starr predicts a 1.6-fold increase of fossil fuel use for electricity generation. Malpas predicts that if the current pace of world demand continues without harnessing the benefits of new technologies, it will be necessary to triple the consumption of coal to satisfy the projected primary energy demand in the year 2020. The potential consequences of this scenario could be profound.

Taken together, the by-products of energy production, distribution, and consumption define society's single largest environmental loading. In recent years, awareness of this loading has evolved from a focus on point sources of pollution with point effects to distributed sources with distributed effects. As Ruckelshaus notes, "We nearly all have environmental consciousness now, whereas nearly all of us grew up without it." Environmental loading and its consequences occur rather rapidly on the local and even regional scale; the time scale over which significant regional effects develop is on the order of 10 to 30 years (see, for example, Graedel's analyses, this volume). The global accumulation of by-products such as carbon dioxide occurs at a slower rate, doubling in 100 to 150 years, but the time interval between the beginning of this trend (circa 1850) and its possible egregious effects is not known.

This mismatch of time scales and the potential for rapid change presented by the greenhouse issue resemble the conditions characterizing the prehistory of the U.S. oil shocks. The conditions that made the oil price shocks possible were observed for more than 20 years. Beginning in 1950, the United States began to import more petroleum products than it exported. During the 1960s, low oil prices also encouraged rapid growth in the demand for oil in Europe and Japan as oil replaced coal. Clear signals of the significance of this demand trend were broadcast. In the early 1960s, OPEC successfully cooperated to resist a reduction in posted prices of oil and proceeded to increase per-barrel royalty and tax payments. During the Arab-Israeli War of 1967, disruptions in the supply of OPEC oil led to the exhaustion of U.S. excess capacity (at that time). By 1971 the experience and lessons in cooperation of the 1960s encouraged OPEC to determine oil production and prices unilaterally, and the embargo followed.

The scale on which energy management in the United States was practiced did not match well the much slower demand growth and geographic shifts of energy supply. As a result, the system failed to respond and vulnerability accumulated as a slowly growing dependence on imported oil. Curiously, this dependence and its implications were well understood (before October 1973, projections estimated that imports would supply more than half the oil consumed by the United States), but it was not until the price spikes of the 1970s that the system was shocked into a flexible state and the active pursuit of "energy independence" was undertaken.

Just as U.S. dependence on foreign oil grew slowly, so too will effects of climate change induced by greenhouse gases take time to manifest themselves. The increase of greenhouse gases is as well established as the increasing oil deficit was in the 1950s and 1960s, and the possibility that these gases contribute to climate change seems as clear as did the potential for OPEC disruptions in oil supply. The complexity of mechanisms by which greenhouse gases may affect climate complicates the interpretation of environmental signals. A few observers speculate that the U.S. drought of 1988 was a signal that the calamity has begun; however, the weight of evidence is that this proposition is false. This complexity makes predicting the future climate as difficult as predicting future actions of the OPEC cartel. From these perspectives, the similarity between the management challenges presented by the oil situation of the 1960s and by the greenhouse situation of today is noteworthy.

The key issue here is the rate of change relative to the rate at which the system can adapt. Thomas Schelling points out that we routinely adapt to rapid climate differences far greater than those predicted by current models of the greenhouse effect on a seasonal and even daily basis; he therefore concludes that societal adaptation, rather than intervention, will result. However, Helm and Schneider argue that current models predict that a transition to a different climate could occur much faster than the rate at which global ecosystems can adapt. If this occurs, a state of unpredictable planetary disequilibrium would result, and this in turn would most likely lead to societal disruption. Because climate system modeling is still approximate, the potential rate of climate change is not known with sufficient precision to predict that disequilibrium will occur. Moreover, social responses to disequilibrium conditions are also difficult to predict.

Weinberg gives further evidence of the importance of relative rates with his observation that energy demand is inelastic in the short term but elastic over the long term; Gibbons and Blair make the same observation in the context of energy efficiency. Thus, should a climate shock occur, responding in a reactive rather than anticipatory manner is likely to be inefficient, and the tremendous inertia of the global climate may make reversal of change impossible.

In sum, like U.S. oil import vulnerability, the potential for climate shocks accumulates slowly. The nature of possible future climates is uncertain, as evidenced by the range of opinion represented in this volume. Helm and Schneider assert that a very rapid and possibly unpleasant climate transition—a greenhouse shock—seems possible, while Schelling argues that the magnitude of currently projected climate change is smaller than changes with which society routinely copes. Together these views constitute a quandary; to assume that the greenhouse situation is benign, only to be surprised later by a "greenhouse shock," would be at least as imprudent as overreacting to normal fluctuations in climate. Despite all that is familiar, this management challenge is unprecedented: The rate of accumulated vulnerability is the slowest yet; the prospect for undesirable change has the longest process horizon yet encountered; and the widest range of cultures, societal goals, and value systems is affected. Further, unlike other societal by-products, greenhouse gases are the by-product of many activities essential to life in all societies. Finally, contrary to the U.S. experience with the OPEC shocks, the effects of a greenhouse shock may be irreversible. Therefore, quite unlike past U.S. efforts at managing the domestic energy system, the world may possess only one chance to prevent a greenhouse shock. Because the ways in which energy is produced, distributed, and consumed contribute the single largest loading of greenhouse gases to the planetary environment, all countries must now take this into account in future energy planning. Thus, in addition to reducing U.S. vulnerability to foreign energy sources, if the United States wishes to contribute to reducing the world's vulnerability to a possible future greenhouse shock, broad reappraisal of the energy system from these perspectives is required.

THE ENERGY PLANNING WINDOW

Because of the environmental dimension of energy production and use, the importance of conducting an energy reappraisal grows in proportion to world population. But considering the cost of changing energy systems and the uncertainties of the environmental interactions, why is it not prudent to postpone a reappraisal pending further study of the uncertainties? The reason is that if the United States is experiencing a period of heightened system flexibility—a window of opportunity—then energy managers must act quickly before this window closes.

Evidence of such an opportunity window is provided in this volume and in the history of energy. In general, the origins of the current system flexibility lie in responses to the energy price shocks of the 1970s. The way in which severe shocks to the system can create technological opportunities is illustrated in the history of energy systems.

During the early 1900s, the United States experienced an episode of energy shortage not unlike that of the 1970s. In 1917 coal supplied 75 percent of the energy consumed by the United States. Then the demands of World War I, mining labor shortages, and railroad tie-ups combined to create a severe coal shortage—within the span of a few months, the cost of coal more than doubled. In immediate response to this crisis, allocation procedures and curtailment of nonessential uses of power were enacted, just as in the emergency of the 1970s. Energy experts also invoked a variety of conservation measures, including fuel adjustments on consumer electricity bills, the extension of daylight-savings time, and lowering of thermostat settings to 68°F (De Simone, 1976). The general coal strike of 1919 caused a second fuel price shock. This strike, the culmination of a series of labor actions that took place after the war, virtually shut down bituminous coal production. However, this chapter in the history of American unions may well have been a Pyrrhic victory, because the price effects felt throughout the energy system appear to have shocked it into a state of flexibility. Petroleum energy, already poised for rapid market penetration, quickly became the dominant source of energy supply. Once the demand pressure for energy resumed in the late 1920s, new uses of coal, such as conversion into liquid fuels, never achieved commercial success.

In World War I, oil revolutionized warfare; fast ships, planes, and trucks were all decisive contributors to victory. The new possibilities that these machines and technologies offered led to a rapid increase in the demand for oil. In fact, the United States became a net oil importer from 1920 to 1922. Understandably, the consensus among energy experts at that time was that the United States did not possess an adequate supply of petroleum and that somehow a future of continued industrial progress must be ensured though a sustainable reliance on coal. At that juncture, the two most actively discussed energy options were the development of large-scale electric power systems and finding petroleum substitutes; making oil from coal was a popular contender at the time. With the exception of nuclear power, almost every alternative energy technology explored in the 1970s had been envisioned by the studies of the 1920s.

Returning to the present situation, the current flexibility in the energy system results from the price shocks of the 1970s. Because many efforts over the past 15 years toward developing a more stable and robust configuration of the system were premised on relatively high long-term oil prices, they were unsuccessful. Hence, even though the OPEC price shocks occurred several years ago, the energy system continues to be more flexible and exploratory than at most other times in the recent past. Yet, as Behnke and others observe, the growth rate of energy demand is resuming. Although the current supply system has been able to respond and maintain low prices, it has done so in reactive rather than anticipatory ways. These measures have included purchasing electric power rather than building new generating capacity, adding any new capacity in small incremental steps, expanding the flexibility and throughput of current facilities such as refineries, and moving to fuel flexibility for stationary applications. Ironically, as Gibbons and Blair note, the value of efficiency in this regard seems to have fallen, as evidenced by the recent debate over relaxation of the corporate average fuel economy (CAFE) standards for automobiles and efficiency-promoting speed limits. Ruckelshaus finds this state of affairs paradoxical, for it is during periods of low energy prices that "more resources are available to make conservation investments against the inevitable day when the price of energy goes up again." He concludes that this time of low energy costs must be used as a grace period. This period will end when demand growth resumes at a rate beyond the capacity of the current system to add supply. Under such pressure, the energy system will not be able to assume the technological flexibility needed to incorporate fundamental, as opposed to incremental, improvements.

THE ROLE OF TECHNOLOGY

The positive role of technology is noted by most of the authors in this volume. Starr identifies the many present, past, and future possible surprises made possible by electric power. New technologies using lasers, plasma torches, superconductors, and advanced materials can lead to processes based on organic-plasma chemistry, new electrochemistry (including batteries), and new methods for materials processing and fabrication. If the demand anticipated by Starr materializes, a number of promising technologies discussed by Behnke may play an important role. These technologies include computers, telecommunications, microelectronics, and high-power semiconductor devices such as thyristors. Bookout also notes how new exploration, production, and refining technology has made important contributions to our domestic petroleum supply, and Henrick Ager-Hanssen predicts that technology will continue to play a corresponding role for natural gas supply. Some of the new technologies that have contributed to expanding our hydrocarbon resources include digital processing and enhancement of seismic and remotely sensed data and drilling in harsh conditions on offshore drilling platforms. New technologies that use natural gas for power generation, such as gas turbines and boiler cofiring, are described by William McCormick. The role of technology in achieving efficiency is apparent from the discussion of Gibbons and Blair. Technology has improved the efficiency of heating, illuminating, air conditioning, cooking, transportation, appliances, and industrial processes; its potential is far from exhausted. Malpas asserts that technology will be the primary agent for continued efficiency in the future.

It is the historical and potential role of technology that gives these authors their pragmatic optimism. Clearly, technology has been a primary tool for both developing new energy sources and using them more efficiently. So while technological surprise may limit visibility into the future, it clearly makes possible brighter futures than those that presuppose a Malthusian world of exhausted nonrenewable resources. Ultimately, mankind may reach a planetary limit to growth, in which case, technology will be crucial to getting there smoothly.

In principle a wide variety of developed technologies can be employed to meet a growing demand, but in practice there are lead times inherent in implementing any new energy strategy. Furthermore, several important supply issues may be confronted soon: for example, the need for more electrical generating capacity and the possible end of the gas deliverability "bubble." Together these factors may cause the grace period of low U.S. energy prices to end during the 1990s. If so, the option to implement fundamental improvements is likely to become too costly, and only enhancements and extensions to the existing system will be possible. In this light, the decade of the 1990s becomes an especially important period for decision making and planning.

SOME IMPLICATIONS FOR STRATEGY

If past energy appraisals and forecasts were flawed because they did not anticipate the complexity of the future, why is this not an inherent limitation of any new reappraisal? It seems inevitable that some future issues will not be discovered in forthcoming reappraisals. The answer is that, if nothing else, the recently recognized economic and environmental issues on regional, national, and global scales have yet to be incorporated into energy system modeling and planning. More than ever, technology and fuel choices must factor environmental consequences and social trade-offs between benefits and costs.

The present constrained circumstances for investment in a new energy system present an enormous challenge to domestic energy institutions. There is increasing indication that the risks of environmental and price-supply "future shocks" need to be anticipated with insurance strategies soon. But, how do planners formulate the first of what will doubtless be a series of "midcourse corrections" in the evolution of energy systems? What should be chosen as the rational first steps of departure from the status quo to promote new increments of energy supply that are both affordable and more environmentally benign? Which of these steps will be mandated by policy, and which will be the fruits of a market system producing alternatives by economic incentives?

From the above discussion, we conclude that energy policies and choices must be resilient to surprises, both good and bad. This resiliency must pervade all sectors of the planning establishment: business, municipalities, utilities, and governments. Although the particulars of laws, regulations, taxes, and national programs must come from other forums, the following clear and promising directions for energy policy emerge from the presentations in this volume:

1. Increase energy efficiency.

2. Move the aggregate fossil fuel mix toward a higher hydrogen content.
   3. Revisit "smokeless" energy sources, primarily improved nuclear fission.

Policy directions for other important issues, such as transregional acid deposition and local ozone nonattainment, as indicated in those chapters that specifically discuss them, are already being pursued, because the technological needs to address them are visible and the political process to solve them is already in progress.

Increasing Energy Efficiency

Nearly all of the authors in this volume advocate increased energy efficiency as a major element of any future energy plan. The special relevance of this strategy is underscored by the diversity of perspectives and management contexts represented by these individuals. But although the value of increasing efficiency is clearly identified as an important agenda item, the means of achieving gains comparable to those of the past are not. The petroleum price shocks of the 1970s and 1980s drove U.S. energy institutions to perform the easier improvements. For example, some of these gains were based on the improved design of long-life capital stock such as automobiles, buildings, appliances, and factories; the considerable impact of these capital improvements will not be easy to repeat. Although the limits to additional progress are not known, the record of technological advances offers hope that the progress to date is a technological point of departure for considerable additional gains. The planning question is concerned less with identifying potential improvements in technology and design than it is with how to bring them into being.

History suggests that the price of energy usually drives this progress. The higher the price, whether by taxation or market mechanisms, the more the incentive for energy efficiency. The option may be thought of as a trade-off between moderate, scheduled price increases now and possible large shocks in market prices later. However, because of the large energy component in most manufactured and agricultural products, the social cost of price-induced efficiency in these products would most likely be an increment of general inflation and an erosion of the global competitiveness of those nations that do not balance the factors influencing energy supply and demand in the most cost-effective way. An alternative proposal could be to increase the efficiencies mandated for automobiles, appliances, buildings, and residences. This strategy is actively becoming ensnared by the social trade-off process, as witnessed by the recent CAFE standards debate. The design of any energy saving strategy is likely to be controversial because some form of government intervention, be it taxes, sheltered prices, or mandated performance, will almost surely be required. As many of the authors assert, efficiency improvement strategy that is politically attainable will be worth the effort.

Promoting Hydrogen-Rich Fuels

Increasing the use of hydrogen-rich fuels is simple in concept: namely, increase the hydrogen-to-carbon ratio of the aggregate fossil fuel mix. Hydrogen burns to yield energy and water vapor, whereas carbon burns to yield energy and carbon dioxide, a greenhouse gas. The environmentally most desirable hydrocarbons from the standpoint of the greenhouse effect would involve as much hydrogen and as little carbon as possible. Of the currently available fuels, natural gas (methane) scores highest on this value scale, and coal, being mostly carbon, scores lowest. Intermediate choices, in order of decreasing desirability, are propane/butane and methanol, gasoline, and diesel fuel. In a world that honored this ranking, natural gas would be used as often as possible and the use of coal would be discouraged.

This consideration raises grave doubts about an earlier conceptual strategy to use coal as a bridging, transitional fuel between the petroleum era and some future nonfossil era. A better strategy would replace coal with natural gas. Coal has been considered for decades as a feed stock for making methane, but the cost-effectiveness of that process has been disappointing to date. Further, to avoid exacerbating the environmental consequences of the coal-to-methane conversion process, either huge amounts of nonfossil hydrogen or a non-CO_2-emitting method of excess carbon disposal, will be required. There appear to be abundant world resources of gas, but because it is costly to transport for distances of more than about 2,000 miles, the relationship of producing locations to consuming locations must be considered. Moreover, because methane is also a greenhouse gas, the importance of methane leakage must be assessed. There may be enough gas resource on the North American continent (McCormick, this volume) and Europe (Ager-Hanssen, this volume) for a substantial gas policy initiative, especially if exploratory drilling technology improves with gas price stimulation. A second wave of world gas supply will be available when technology helps find more gas and solves the problem of bringing remote gas to consumers in a cost-effective way. Liquefying natural gas is at present accomplished by a relatively costly technology; however, remote gas could be converted to transportable liquid methanol, although this process does not yet have adequately sized downstream markets. Clearly mobilizing more world gas with cost-effective technology is a most promising fossil fuel strategy. This technological challenge is formidable but has great potential.

Revisiting Nuclear Fission

As Malpas asserts, "Nuclear energy is the cleanest of all . . . . One is surprised that environmentalists do not promote it, demanding that it be made safer than it already is." Much has been learned about the design, construction, and operation of nuclear fission power plants. New, inherently safe, second-generation reactor concepts have been researched and designed (Faltermayer, 1988); some have been tested and demonstrated on a large scale. The inherent safety of these new reactors derives solely from the laws of nature, making them virtually immune to operator error or subsystem failure. Although characteristics differ, all these reactors achieve their safety without human or control system intervention; one type of reactor can withstand a total loss of coolant such as would result from a major structural failure.

The nuclear power plants in the United States today rely on engineered safety systems and possess little immunity to many modes of failure, particularly those that lead to a loss of coolant. The need for the engineered safety systems requires costly and redundant controls to assert a positive, corrective response to subsystem malfunction or operator error. Even though the U.S. nuclear power plants of today have demonstrated the capability of their engineered safety systems to protect the health and safety of the public, experience has shown that they are less successful at protecting the financial risk of their owners.

In addition to inherent safety, some second-generation reactors offer other attractive features such as higher thermodynamic performance and fuel cycle flexibility. For the above and other reasons beyond the scope of this chapter, the inherent safety of the new, second-generation reactors translates to substantially smaller social and financial risk than previously possible.

Weinberg, Starr, McCormick, Balzhiser, and Malpas cite the growing importance of "smokeless" nuclear technology. The promise of inherently safe reactors further suggests that it is time to reconsider and debate anew the nuclear option. However, even though the reactor technology portion is within reach, the balance of the system is not. One important difference is the apparent uneven quality in management. That some U.S. nuclear plants perform as well as any in the world while others are among the worst is widely attributed to this fact (Ahearne, 1986; Hansen et al., 1989); thus, any nuclear reappraisal must include the management dimension. Also, a complete reappraisal must circumscribe the entire fuel cycle and include a socially and politically acceptable solution to the waste problem. Failure to do this in the United States has cost much in terms of economic opportunities in the past and will continue to be costly in terms of energy deficits in the future. Balzhiser offers his view of what nuclear nonpolicy has cost the United States, and one view of what will be required to begin to form an enlightened and constructive nuclear policy. We believe that it is time to stop paying for nuclear nonpolicy; it is time to make the hard choice to either fix or forget nuclear power.

SYNERGIES AND CONCLUSION

Most of the options discussed above are not mutually exclusive; indeed, some of them can be combined synergistically. For example, hydrogen-rich fuels are also those fuels most conducive to reliable, high-efficiency technologies such as combined-cycle gas turbines. If the energy system is to continue to evolve toward fuels ever richer in hydrogen, non-carbon-based hydrogen sources will be necessary. Nuclear energy provides the ultimate means to provide this smokeless energy in large amounts. Further, high-value "energy currencies" such as hydrogen-rich fuels and electricity can be used to run greenhouse-benign transportation systems, such as new-generation trains or electric cars. Abundant electricity can promote system efficiencies and also simplify the substitution of smokeless energy sources.

These three directions, energy efficiency, the hydrogen-rich fuels, and acceptable nuclear power, are emerging in energy policy discussions because they are technically and logically relevant to modern society's energy vulnerabilities. The technical case for them and the abundance of constructive technological surprises implicit within them richly deserve investigation and appraisal. The technical choices may not be easy. The greater challenge, however, will be to marshal the necessary social and political will to make progress along whichever technological paths are chosen. To be ready for the inevitable surprises that the future holds, the time to face the challenge of energy planning in a dynamic world is now.

> If it be now, 'tis not to come;
> If it be not to come, it will be now;
> If it be not now, yet it will come:
> The readiness is all. . . .
>
> Shakespeare, *Hamlet*

REFERENCES

Ahearne, J. F. 1986. Three Mile Island and Bhopal: Lessons learned and not learned. Pp. 197–205 in Hazards: Technology and Fairness. Washington, D.C.: National Academy Press.

De Simone, D. 1976. Technology assessment: Where we have been. Pp. 1–4 in Retrospective Technology Assessment—1976, J. A. Tarr, ed. San Francisco, Calif.: San Francisco Press.

Faltermayer, E. 1988. Taking fear out of nuclear power. Fortune 118(August 1):105–118.

Hansen, K., W. Dietmar, E. Beckjord, E., P. Gyftopoulos, M. Golay, and R. Lester. 1989. Making nuclear power work: Lessons from around the world. Technology Review 92(February):31–40.

1
Supply, Demand, and Reappraisal

Energy in Retrospect: Is the Past Prologue?

ALVIN M. WEINBERG

Before the discovery of fission, energy policy was not a central issue in the United States. After all, this country was blessed with enormous reserves of fossil fuel, and one could hardly conceive of a day when the United States would be importing 30 percent of its oil. To be sure, the 1859 discovery of oil in Titusville, Pennsylvania, came just in time to replace whale oil, which was becoming scarce; and before the discovery of the East Texas fields, alternatives such as shale oil were being pursued. By and large, however, the problem of energy in its broadest aspect had not become part of the federal government's agenda.

The discovery of fission, which was widely regarded as the ultimate answer to the problem of energy, focused attention on energy. It was as though, with the solution in hand, we became aware of the problem. Thus, in a 1953 report sponsored by the Atomic Energy Commission (Putnam, 1953), Palmer Putnam argued that a prudent custodian of the world's energy future should assume energy demand would grow exponentially and that energy supply would turn out to be lower than the expansive estimates of supply then current. Although Putnam's maximum plausible world population by 2050 was only 6 billion people, his per capita growth rate for energy of approximately 3 percent per year was very high; this led to his "maximum plausible" annual demand for energy of 436 quads (1 quad = 1 quadrillion, or 10^{15}, Btu) by 2000 A.D. and 2,650 quads by 2050! No wonder Putnam concluded that the world must get on with the development of all energy sources, especially nuclear power and solar energy (which he regarded as too expensive), as well as improving the efficiency of energy use. Incidentally, Putnam was the first energy futurologist to call attention to the implication of the greenhouse effect for energy policy.

Putnam's report echoed in apocalyptic tone the Paley Commission Report of 1952 (President's Material Policy Commission, 1952). This too warned that serious shortages in energy supplies could develop, but by and large Paley went unheeded. The 1950s and 1960s were periods of energy euphoria, although a few voices, notably King Hubbert of "Hubbert oil bubble" fame, warned that the United States would become an oil importer by the 1970s (Hubbert, 1969).

The euphoria reached its zenith with the 1962 Atomic Energy Commission (AEC) Report to the President on Civilian Nuclear Power, (U.S. AEC, 1962) projecting some 734 gigawatts of electricity from nuclear power by 2000 (this represented about 30 percent of a total projected energy demand of 135 quads), and the 1964 interagency study of energy research and development (R&D) which placed fission into a broader context of energy sources. This report found "no ground for serious concern that the nation is using up any of its stocks of fossil fuel too rapidly; rather there is the suspicion that we are using them up too slowly . . . we are concerned for the day when the value of untapped fossil-fuel resources might have tumbled . . . and the nation will regret that it did not make greater use of these stocks when they were still precious" (Cambel, 1964). Despite this rosy estimate of the U.S. energy future, the interagency committee urged that the government expand research on long-range energy sources, both nuclear and nonnuclear.

These studies belong to what could be called the "pre-Cambrian" period of energy policy. During this period an overall energy policy hardly seemed very relevant; and as for government-sponsored energy R&D, this was almost entirely preempted by the all-powerful AEC and the Joint Committee on Atomic Energy. Although an Office of Coal Research had been set up in 1971, nuclear energy strongly dominated the government's thinking about the future of energy.

President Nixon's price freeze in 1971, followed by the Arab oil embargo in 1973, marked the beginning of the modern era of energy policy. The United States was then importing 6 million barrels of oil per day, and independence soon became the aim of U.S. energy policy. Thus, Dixy Lee Ray (1973), chairman of the AEC, reported to the President in 1973 that the nation could achieve energy independence by 1985—but only if it conserved the equivalent of 14 quads (an oil equivalent of 7 million barrels per day) out of a total annual demand of 100 quads; and Project Independence (1974) claimed the United States could achieve energy self-sufficiency by 1985 at an annual energy consumption of 96.3 quads if oil prices rose by 20 percent and the nation conserved approximately 8 quads.

These estimates of future demand were on the low side. Most forecasters at the time were predicting a 1985 energy demand of around 115 quads. Even Amory Lovins (1977), an arch-exponent of limited growth, was predicting 90 quads—a number close to the Ford Foundation's "Zero Growth" scenario (Energy Policy Project, 1974). Only the National Research Council's Committee on Nuclear and Alternative Energy Systems (CONAES; Brooks and Hollander, 1979), in its heavy conservation scenarios, spoke about the demand for energy remaining constant—or even falling—but CONAES characterized its extreme conservation scenario as "very aggressive, deliberately arrived at reduced demand requiring some life-style changes" (see scenario A* in Gibbons and Blair, Figure 5, in this volume). I do not think that CONAES took this scenario very seriously.

In those days, several energy analysts used as a rule of thumb that the number of quads equaled the last two digits of the calendar year—78 quads in 1978, 79 in 1979, and so on. However, the reality turned out very differently. Who, in 1973, would have predicted that the total amount of energy used in 1986 would be only 74 quads, the same as in 1973? Let energy forecasters practice their precarious art with humility!

THE RATIO OF ENERGY TO GROSS NATIONAL PRODUCT

In the early 1970s many of us were convinced that the ratio of energy (E) to gross national product (GNP) or to gross domestic product (GDP) was a constant—as indeed it was from 1945 to 1975. We seem to have forgotten that the E/GNP ratio had been falling from 1920 to 1940. The constancy of this ratio from 1940 to 1970 concealed the secular trend toward higher efficiency (Figure 1). This improvement in energy efficiency was evident throughout the Organization for Economic Cooperation and Development (OECD): between 1966 and 1970, the elasticity of energy to gross domestic product, $\epsilon \approx \frac{\Delta \ln E}{\Delta \ln GDP} \approx 1.4$; between 1980 and 1984, $\epsilon \approx -0.2$ (see Table 1). Although the energy demand in the less developed and newly industrialized countries continued to expand, the entire noncommunist world became considerably more energy efficient: $\epsilon \approx 1.3$ in 1966–1970; $\epsilon \approx -0.5$ in 1980–1983.

The other characteristic trend has been the continued electrification of the United States and of the world. In 1968, some 18 percent of primary energy in the United States was converted to electricity; by 1987 this fraction had doubled. The figures for the noncommunist world are similar. Moreover, the elasticity ratio of electricity to GNP seems to have been fairly constant, at least for the past 40 years (Figure 2).

Thus, the great realities of energy in the postembargo world have been (1) the extraordinary flattening in the demand for energy in the developed world, which implies an unexpected decoupling of energy and GNP, and (2)

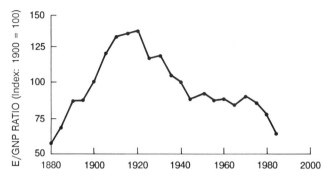

FIGURE 1 Energy (E; mineral fuels and hydropower) consumed per dollar of real gross national product (GNP) in the United States, 1880-1985. SOURCE: Schurr (1984:410).

the continuing electrification of the world in conjuction with the remarkable correlation between electricity and GNP.

How can this extraordinary diminution in the growth of energy demand be explained—a diminution so extreme as to call into question the usefulness of forecasting energy demand? Four schools of thought have arisen to explain away this discrepancy.

First are the *energy economists*. For them, the reduction in energy demand simply reflects both the lowered rate of economic growth and the increase in the price of energy. Even today, the average price of all energy is some two times (in real terms) the price of energy 15 years ago—little

TABLE 1 Growth Rates of the Demand for Energy and the Gross National Product (GNP) of the Noncommunist World

	1966–1970	1970–1975	1975–1980	1980–1985
Organization for Economic Cooperation and Development				
Energy (%)	5.8	1.3	1.7	−0.2
Gross domestic product (%)	4.2	3.0	3.5	2.1 (to 1984)
Elasticity	1.4	0.4	0.5	−0.2 (to 1984)
Noncommunist world				
Energy (%)	6.0	1.9	2.6	0.7
Gross domestic product (%)	4.5	3.6	3.9	1.4 (to 1983)
Elasticity	1.3	0.5	0.7	−0.5 (to 1983)

SOURCE: BP statistics—International Monetary Fund–*International Financial Statistics* as quoted in Ministry of International Trade and Industry (no date), p. 59.

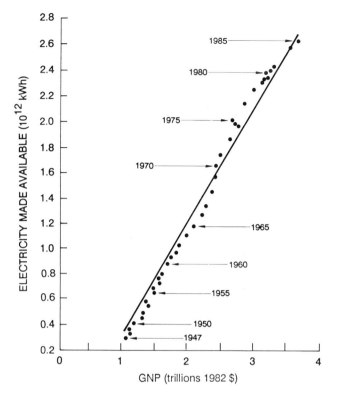

FIGURE 2 Electricity production in the United States versus gross national product (GNP). The approximately linear relationship suggests that the elasticity ratio of electricity to GNP has been constant since 1947. SOURCE: Energy Study Center, Electric Power Research Institute.

wonder that demand has abated! Moreover, since all this has resulted from the operation of the market, energy economists by and large support nonintervention as our basic energy policy. The market has worked; let it continue to work.

A second group of analysts is the *structuralists*. For them an important, if not dominant, reason for energy being uncoupled from GNP is a change in the structure of our economic activity—the shift from manufacturing, mining, and agriculture to services, coupled with a saturation in some end uses (e.g., television sets and refrigerators). Because services are by and large less energy intensive than manufacturing, energy demand has flattened. Thus, the structuralists' policy implication would be: encourage further shift to services; energy will then take care of itself.

Third are the *conservationists*. These include the doctrinal conservationists who regard conservation of energy as a transcendent human purpose; and the technical conservationists who simply insist that the technology of efficiency, both in end use and in energy production, has improved greatly and can be improved further. Moreover, these improvements often result in lower overall cost. For them much of the reduced E/GNP ratio reflects the adoption of more efficient technologies prompted in part by the rise in energy prices, and in part by the widespread acceptance of a conservation ethic. Energy policy must, therefore, stress increased technical efficiency; but—at least for doctrinal conservationists—the government must mandate efficiency standards such as corporate average fuel economy (CAFE) and building efficiency performance standards (BEPS), as well as promote broad acceptance of the principle that conservation per se is ethically superior to any alternative.

Finally, there are the *"electro-niks."* For them the increase in electrification and the reduction in E/GNP are not coincidental. Rather, electrification of industry is per se a powerful catalyst of increased productivity. As industry electrifies, whether or not the electrification is itself energy efficient, industry becomes more productive. Electrification increases the denominator of the E/GNP ratio rather than diminishing the numerator, but the result is a decoupling of energy growth and economic growth. Energy policy for the electro-niks is: encourage electrification because an electrified society is an energy-efficient society.

There is truth in the views of all four groups of analysts. Nevertheless, although analysis of energy supply and demand is much more sophisticated now than it was at the time of Project Independence, all of us must retain a basic skepticism about the ability to predict, much less mold, energy futures even to the year 2000. An intrinsic dilemma confronts us: Energy policy, insofar as it requires decisions today that affect the world 10, 20, even 50 years hence, must rely on visualization of that future world; yet as the inglorious history of past energy projections has demonstrated, that world cannot be known. The main lesson from this experience is that, insofar as possible, we must try to formulate energy policies that finesse these uncertainties, that are resilient to surprises. Is this a realistic possibility?

INCREASING SUPPLY AND REDUCING DEMAND

Balancing supply and demand is of course automatic. The issue is how to achieve this balance without causing unacceptable economic and social dislocations. Most old-time energy people assumed almost automatically that demand was hardly subject to any control. Their emphasis, both as policymakers and as engineers, was on increasing supply. Develop nuclear fission and fusion, oil shale, synfuels, geothermal sources—even solar and wind—was their response to the oil embargo.

What a shock to discover that demand could also be altered and, indeed, that the technologies required to reduce demand might challenge the engineering community no less than the technologies of expanding supply. Today, there are many opportunities for research in demand management as well as in supply enhancement.

Why did most engineers gravitate toward supply enhancement rather than demand management? I can give at least two reasons. First, designing nuclear reactors seemed to be more glamorous than improving car efficiencies. Second, demand management usually requires millions of people to change their way of doing something. In some cases, such as lowering the thermostat, the change might affect life-style; in other cases, such as replacing an energy-inefficient refrigerator with a more efficient one, the change requires an additional outlay of money. In a broad sense, demand management, even when based on clever new technologies, is a social fix: many individual decisions are needed to achieve lower demand. By contrast, increasing supply was regarded, perhaps naively, as a purely technical fix: only a few people have to be convinced to build a nuclear reactor or a synfuel plant. The technical fix seemed to be simpler than the social fix.

At least, this is the way many engineers viewed the matter. Somehow, predicting the increased supply seemed much more robust than predicting reduced demand. But we were wrong again. We neglected the public antagonism to all sorts of large centralized energy systems—whether nuclear reactors, coal-fired power stations, or synfuel plants. We also underestimated the public's acceptance of conservation—whether price induced or resulting from a widespread belief in a conservation ethic reinforced by government-mandated efficiency standards.

In a way, the relative effectiveness of supply enhancement and demand management reflects the underlying U.S. political structure. Ours is a Jeffersonian democracy: decentralized, open, some would say chaotic. Large-scale interventions that are perceived as threatening by a determined group can be, and often are, blocked. Under the circumstances, we have much incentive to avoid big, threatening, energy supply projects in favor of demand management and much smaller, decentralized, supply options.

By contrast, where the political structure is elitist, particularly in France with its Jacobin political tradition, large, centralized supply options—especially nuclear energy—remain viable. France has managed to reduce oil imports (and, incidentally, reduce the carbon dioxide thrown into the atmosphere) largely by its steadfast commitment to nuclear electrification—a path that is unavailable to the United States, at least for the present.

INCREMENTALISM AND ITS CONSEQUENCES

In trying to formulate energy strategy for the next decade, we must therefore accept three realities:

1. The future is much less knowable than it was thought to be 15 years ago.
2. Ours is a participatory polity, one with growing environmental concern.
3. Although certain segments of the U.S. energy supply system, notably oil, are dominated by a few large corporations, other segments, particularly electricity, are fragmented. The United States has approximately one generating company per million people, whereas Japan has one per 10 million and France, one per 50 million!

The U.S. energy system seems to be responding to these realities with a grand strategy that I would describe as *incrementalism*. Because our energy demand in 10 years cannot be predicted, we should not build anything very large. Thousand-megawatt power plants or 100,000-barrel-per-day synfuel plants are much too risky. If more electricity is needed, build a 50- or 100-megawatt gas turbine; buy electricity from small independent producers (who have enjoyed the protection of the Public Utilities Regulatory Policy Act of 1978) or, if possible, from Canada, and do not shut down old plants; or reduce demand by offering incentives to customers to use more efficient devices. Such incrementalism finesses the uncertain future and does not expose a generating company to the risk of bankruptcy, which has engulfed unlucky utilities saddled with ridiculously overpriced nuclear reactors. Incrementalism also evokes little antagonism from politically sensitive, and often powerful, conservationists. Thus, the 1950s vision of a gradually nuclear electrified nation has in the 1980s and 1990s given way to a nation in which conservation is primary and in which energy is increasingly supplied by small, decentralized units or by older units that have been coaxed into a few more years of operation. The energy dream of 1950 is coming to pass, but not in the United States. It is happening in France, in Japan, in several communist countries, and in other newly industrialized countries whose political tradition is more authoritarian than ours and whose energy systems are more centralized.

Let us concede that incrementalism is inevitable during the next decade or so, at least if conservation is insufficient to keep our energy demand from growing. Are there dangers in the long run for an energy system that will eventually be dominated by a large number of small producers?

I see several such dangers. Perhaps most important, many of the new electrical supply increments use gas- or oil-fired turbines. Although the Gas Research Institute (GRI, 1987) has recently estimated that there will be enough gas through 2010 (see Table 2), one is naturally concerned

TABLE 2 Gas Research Institute Baseline Projection of Gas Supplies (quads), 1987

Production Basis	1986	1990	2000	2010
Current practice				
Domestic production	16.6	16.3	14.3	9.3
Canadian imports	0.8	0.9	1.4	1.1
Liquefied natural gas imports	0.0[a]	0.1	0.3	0.8
Supplemental sources	0.0[b]	0.2	0.3	0.3
Total	17.4	17.5	16.3	11.5
New initiatives				
Lower-48 advanced technologies	0.0	0.3	2.6	5.4
Alaskan pipeline	0.0	0.0	0.0	1.2
Canadian frontier	0.0	0.0	0.5	0.7
Other imports	0.0	0.0	0.0	1.0
Synthetics	0.0	0.0	0.0	0.1
Total	0.0	0.3	3.1	8.4
Total supply	17.4	17.8	19.4	19.9

[a] Less than 0.05 quad.
[b] Includes net injections to storage of 0.1 quad.

SOURCE: Gas Research Institute (1987).

that to achieve its 19.9 quads of gas projected for 2010, some 8.4 quads must come from "new initiatives" such as advanced extraction technologies, the Alaskan pipeline, the Canadian frontier, imports, and synthetics; and the total imports (including Canadian imports) amount to 3.6 quads. Also, insofar as oil is used in these small generators, U.S. dependence on imported energy will be increasing, not decreasing.

Second is the economy of scale. During the era of energy euphoria, particularly nuclear euphoria, bigger was assumed to be cheaper. The catastrophic escalation of capital costs for nuclear plants has dissuaded us from this belief: we seem now to believe that the economic scaling laws have been repealed or at least can be circumvented if the devices are manufactured serially. However, even if the capital costs of small units are favorable, most would claim that operating costs will be higher for small plants than for large plants.

One must therefore ask, does the trend toward incrementalism imply that energy—particularly electrical energy—will always be more expensive in the United States than it is in countries where large plants continue to dominate?

Although I am not optimistic on this score for the next decade, I see some hope for diminution in the price of energy in the trend toward extending the life of power plants and other large supply devices. The 30-year licensing lifetime of nuclear power plants was a relic of fossil-fueled tradition; fuel efficiency increased at a rate that made power from 30-year-old plants more expensive than from newly built plants. However, with efficiencies plateauing, or being rather irrelevant in the case of nuclear power plants, the incentive to shut down old plants has weakened. Extending the life of an old plant, whether nuclear or fossil, is often cheaper than building a new one. In addition, if our energy system is dominated by plants that have already been paid off and have low operating and fuel costs, the price of energy may once more begin to fall.

This phenomenon may not be confined to electric generators (Weinberg, 1985) but may be applicable to synfuel plants or even solar electric systems, provided that operating costs are low. Thus, a synfuel plant, which might cost $100,000 per daily barrel ($330 per annual barrel) and uses coal at $40 per ton, will produce synfuel at about $92 per barrel; of this, $66 per barrel is capital cost (at 20 percent). However, if the plant lasts a century, rather than the 30 years over which it is amortized, and if its maintenance costs can be kept low, the cost of the synfuel falls to around $25 per barrel once the plant has been amortized. Thus, the first South African Coal, Oil, and Gas Corporation (SASOL) plant, which was placed in operation almost 30 years ago and is now presumably amortized, probably produces synfuel at costs close to the world spot price of oil.

In constructing and modifying the energy system, we must recognize that the energy system is one of society's most basic infrastructures, and—like most infrastructures—it affects not only this generation but future generations as well. Perhaps the moral to be drawn is that future precepts of engineering design ought to stress longevity of the energy-producing device, more so than in the past. Although we may not succeed in giving our own generation the gift of cheap energy, perhaps we will be providing this gift to our children's children.

INTERNATIONAL PERSPECTIVES

My viewpoint has been primarily American and possibly nuclear, electrical, and supply oriented. Yet it seems fair to say that U.S. energy policy during the coming decade will be demand dominated—that the emphasis will be on trying to reduce demand. In this the United States is joined, for example, by Japan; the recent Ministry of International Trade and Industry (MITI, undated) report begins with an analysis of how well demand can be managed. Yet even MITI's minimum-demand scenario projects total energy in Japan expanding at 1.6 percent per year until 2000, and 0.8 percent

TABLE 3 Forecasts of Energy Demand (in million barrels of oil equivalent per day [mboe/d])[a]

	1980 Actual Results	Study Results				Goldemberg et al.
		IASA[b]		WEC[c]		
Estimated year	—	2030		2020		2020
Population (billion)	4.43	7.98		7.72		6.95
		High	Low	High	Low	
Growth rate per capita of GDP[d] (% per year)		2.1	1.1	2.0	1.1	—
Primary energy requirements						
World (mboe/d)	144.4	497.0	309.2	348.8	271.1	158.2
Industrial nations	98.8	283.8	190.6	209.0	176.5	55.1
Developing countries	45.6	213.2	118.6	139.8	94.6	103.1

[a] Original values expressed in terawatts (TW) were converted into million barrels of oil equivalent per day (mboe/d); 1 TW = 14.12 mboe/d.
[b] Analyses from the International Institute for Applied Systems Analysis (1981).
[c] World Energy Conference (1983).
[d] Gross domestic product.

SOURCE: Goldemberg et al. (1985:622).

per year from 2000 to 2030, compared with a growth rate of 1.3 percent per year from 1975 to 1985. Thus, MITI expects the total energy demand in Japan to be at least 62 percent higher in 2030 than it is today and in its maximum-demand scenario 211 percent higher. Nor is Japan attracted to U.S. incrementalism. It expects to continue to build 1,200-megawatt reactors, even as it diversifies supply.

Looking at the entire world, one must be struck by the developing countries' economic growth and increasing use of energy. But the diversity of estimates for the future remains (Table 3): from the World Energy Conference's high scenario of 735 quads in 2020 (World Energy Conference, 1983) to the Goldemberg et al. (1985) estimate of 320 quads in 2020—about the same as today's 300 quads. Everyone seems to agree that the developing countries will use a larger fraction of the world's energy than they now do, but there is little agreement as to how much the total is going to be. Whether this presages a big spurt in oil prices—perhaps even a 1973- or 1979-type energy crisis during the next 20 years—or a much more gradual increase, no one can say.

CONCLUSIONS

Commonplace, even tiresome, is my observation that forecasts were all wrong, that demand has ameliorated, and that we do not know what is going to happen in the next decade or two. I have argued here that under such a veil of uncertainty we can only try to choose policies, technological options, and R&D strategies that are as resilient as possible to surprises. This suggests that the coming energy decade will be the decade of creeping incrementalism: Initiatives, whether enhancement of supply or diminution of demand, will be small and, in the short run, low risk.

At least, we have discovered what has not worked, in particular that the magical talisman of nuclear energy simply was neither magical nor a talisman. It faltered because nuclear optimists ignored social, political, and economic realities.

Perhaps what we have learned most of all is that the market is more powerful than government intervention. This was most strikingly demonstrated by the experience of the Synfuels Corporation. If the energy crisis was the moral equivalent of war, than why did the technique that worked so well in World War II, in the case of synthetic rubber, not work for synfuels? Alas, markets can be circumvented by government ukase in wartime, but not in peacetime. Synthetic oil was just too expensive.

So the U.S. experience with energy since 1973 seems to bear out the views of Frederick Hayek more than those of Maynard Keynes, let alone Karl Marx. During the 1930s, Hayek insisted that government intervention in the economy inevitably would fail because the detailed information, on which the market operates, could never be available to the government interveners. (His debates with Keynes were high points in the history of economic thought.) The past 15 years suggest that, on the whole, governmental interventions have not been very effective. Our fragmented, participatory political structure dooms government energy policies such as those of France or Japan to failure in the United States.

Does this mean that the best policy with respect to energy is to have no policy; that we would do best to dismantle the Department of Energy, stop all tax and direct subsidies, and get the government out of energy policy and energy R&D?

I cannot accept such a conclusion. After all, some government interventions, such as the CAFE standards and home appliance efficiency labeling, probably have helped. Also, although markets are efficient, they are—as Mans Lonnroth points out—myopic; and they lack compassion. Although incrementalism (largely market driven) seems inevitable at present, in the long run it may saddle us with unnecessarily expensive energy. Although nuclear energy is now too expensive and unpalatable to a large minority, if not a majority, of Americans, the incentive to develop inherently safe reactors that will be both economic and acceptable remains strong. In addition, although doctrinal conservationists claim that demand management alone can defeat carbon dioxide, most engineers seem inclined toward an eventual shift, worldwide, to nonfossil energy systems—most probably fission, but perhaps solar and fusion. Thus, a long-term role for government in providing the technical base for those elements of an energy system that are *not* mediated by the market seems proper.

Engineers are instinctively technical fixers; we are suspicious of social fixes, perhaps because we do not understand them, perhaps because we regard them as more difficult than technical fixes. Looking toward an unknowable future, we recognize that despite its social components, our energy future will depend ultimately on ingenuity: on cheap variable-speed motors, practical energy storage devices, more efficient cars, economical photovoltaic systems, inherently safe reactors, and perhaps even successful fusion. The menu of technical challenges is large, much larger than was realized when the Joint Committee on Atomic Energy dominated U.S. government energy R&D policy. A challenge so large and so diverse is what engineers thrive on. The engineering community—perhaps a more sober and more realistic community—must dedicate itself to providing the technical basis for a more rational and resilient energy future.

REFERENCES

Brooks, H., and J. M. Hollander. 1979. United States energy alternatives to 2010 and beyond: The CONAES study. Annual Review of Energy 4:1–70.

Cambel, A. 1964. Energy R&D and National Progress. Interdepartmental Energy Study Group. Washington, D.C.: U.S. Government Printing Office.

Energy Policy Project. 1974. A Time to Choose America's Energy Future. Cambridge, Mass.: Ballinger Press.

Gas Research Institute (GRI). 1987. Gas Research Institute Baseline Projection. Washington, D.C.: GRI.

Goldemberg, J., T. B. Johansson, A. K. N. Reddy, and R. H. Williams. 1985. An end-use oriented global energy strategy. Annual Review of Energy 10:613–688.

Hubbert, M. K. 1969. Energy resources. Pp. 157–242 in Resources and Man. San Francisco: W. H. Freeman.

International Institute for Applied Systems Analysis. 1981. Energy in a Finite World—A Global Systems Analysis. Cambridge, Mass.: Ballinger.

Lovins, A. 1977. Energy strategy, the road not taken? Future Strategies for Energy Development, A Question of Scale. Oak Ridge, Tenn.: Oak Ridge Associated Universities.

Ministry of International Trade and Industry (MITI). Undated. The Twenty-First Century Energy Vision—Entering the Multiple Energy Era. MITI, Japan.

President's Material Policy Commission. 1952. Resources for Freedom, A Report to the President. Washington, D.C.: U.S. Government Printing Office.

Federal Energy Administration. 1974. Project Independence, A Summary. Washington, D.C.: U.S. Government Printing Office.

Putnam, P. 1953. Energy in the Future. New York: Van Nostrand.

Ray, D. L. 1973. The Nation's Energy Future, A Report to Richard M. Nixon, President of the United States. Washington, D.C.: U.S. Government Printing Office.

Schurr, S. H. 1984. Energy use, technological change and productive efficiency: An economic-historical interpretation. Annual Review of Energy 9:410.
U.S. Atomic Energy Commission. 1962. Civilian Nuclear Power: A Report to the President—1962. Washington, D.C.: U.S. Government Printing Office. (Also 1967 Supplement, issued February 1967.)
Weinberg, A. M. 1985. Immortal energy systems and intergenerational justice. Energy Policy 13(1):51–59.
World Energy Conference. 1983. Energy 2000–2020: World Prospects and Regional Stresses, J. R. Frisch, ed. London: Graham and Trotman.

Energy: Production, Consumption, and Consequences. 1990. Pp. 35–51. Washington, D.C.: National Academy Press.

Energy Efficiency: Its Potential and Limits to the Year 2000

JOHN H. GIBBONS AND PETER D. BLAIR

In the late 1960s and early 1970s, researchers observed that energy production as well as the efficiency of energy use was increasing steadily, even in the face of falling energy prices. As a result, many concluded that it might be timely to investigate the nature of demand—where energy goes, how efficiency is limited, and what effect price has on that efficiency. In 1973, one of the authors (JHG) was asked to set up the first energy conservation efforts in the federal government, including public education, federal agency actions, and research support. At the Federal Energy Office, a wide range of energy conservation efforts were initiated hurriedly, including what turned out to be long-term successes such as building and appliance standards, efforts to increase public awareness, research and development initiatives, and some policy initiatives.

During that period, many misconceptions and vivid images about the nature of energy conservation emerged. For example, the image of President Carter wearing a cardigan sweater and seated near a fireplace as he addressed the nation may have stimulated President Reagan's definition of conservation as "being cold in the winter and hot in the summer" and supported the popular view that "the United States didn't conserve its way to greatness, it produced its way to greatness." Since that time, however, conservation has been viewed increasingly as an economically rational—if not imperative—global strategy. Decision makers are now cognizant that both market and nonmarket signals affect the rate of adoption of conservation opportunities. Also, the enormous potential for technological innovation in energy conservation has become more widely appreciated. Because of widespread negative connotations, the term that is used in preference to energy *conservation* is energy *efficiency*, which more readily conjures up the positive images of competitiveness and productivity rather than the negative images of self-denial, retreat, and sacrifice that have been attributed so persistently to conservation.

Over the past decade, the motivation for adopting energy efficiency improvements has changed considerably. The sense of urgency about the security of energy supply is currently not the driving force behind such improvements. Instead, the desire to improve the competitiveness of U.S. products in world markets and the concern over both short- and long-term environmental impacts of burning fossil fuels seem to dominate. The short-term effects of fossil fuels on air quality (e.g., acid rain, ozone), as well as the long-term effects on global warming, are now of increasing concern. This changed decision-making environment will have important impacts on the appropriate policy tools for capitalizing on the benefits of energy efficiency. In particular, beyond the more frequently cited concerns such as economic vulnerability and national security, there is growing evidence that gases emitted during the burning of carbon-rich fossil fuels are substantial contributors to global greenhouse warming. This concern provides a new incentive to reduce the nation's dependence on these fuels.

HISTORICAL PERSPECTIVE

After the oil price shocks of the early 1970s, coincident with the reversal of the marginal cost of electricity (from decreasing to increasing), came several years of intense analysis as well as action. The efficiency of energy use rose rapidly because of improved "housekeeping," retrofits, and a variety of reversible options such as lowering thermostats and speed limits. However, products that use energy more efficiently began to emerge too. The effort of the National Research Council's Committee on Nuclear and Alternative Energy Systems (CONAES) in the mid-1970s was a key activity in the national attempt to understand energy demand. The study (National Research Council, 1979) was important not only in its analysis, but also in drawing together the divergent views of engineers and economists about the dynamics of energy supply and demand.

In the early 1970s many features of energy consumption were explained in detail and many opportunities were developed for markedly increasing energy efficiency. There were also many surprises. For example, Figure 1 shows efficiency as a function of cooling capacity for unit air conditioners in 1973. Note that at the time the price of energy seemed to have little to do with efficiency. Efficiency improvements had been undertaken by manufacturers primarily to allow higher capacity units to be used in standard household electrical circuits. Another perhaps less surprising example was

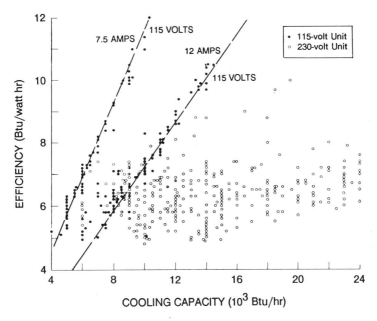

FIGURE 1 Efficiency of room air conditioners. SOURCE: National Research Council (1979).

refrigerators. Figure 2 shows the daily energy use for refrigerators as a function of their retail price. Various design features are shown as they were incorporated, which illustrates the dramatic trade-off between initial capital cost and energy savings as a function of those features.

At the time of the CONAES study, the potential for energy savings in heating and cooling systems was just beginning to emerge. For example, Figure 3 shows the energy balance of an average residential heating system in 1975. Note that the net (heat capture) efficiency for new units in 1975 was 55 to 65 percent. New units today are in the 95 percent range.

A particularly important lesson of CONAES was in transportation. The CONAES scenarios of future energy paths (Figure 4) showed that after initial gains in fuel economy (e.g., up to approximately 20 miles per gallon), the role of fuel became less important in minimizing the total cost of driving. In fact, in all scenarios the marginal gains from additional fuel economy improvements had little effect on the total cost per mile of operation over the lifetime of the car. This may indeed explain why automakers are not investing in new energy efficiency improvements: car buyers are indifferent, partly because other features such as performance are more important to them and partly because further efficiency gains would contribute relatively little to the reduction in total operating costs. This does not mean that

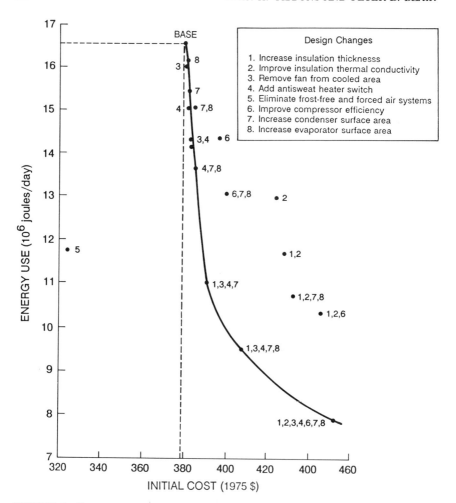

FIGURE 2 Energy use versus retail price for refrigerator design changes. SOURCE: National Research Council (1979).

in the aggregate such savings would not be significant, especially for the nation. On the contrary, they would, but to stimulate adoption, policy intervention is likely to be required to complement the market. Policy intervention will most likely continue to be very difficult to accomplish, however. For example, the 1988 rollback of the corporate average fuel economy standards set by the Environmental Protection Agency (EPA) for new vehicles suggests that some pressures against such policy intervention are increasing.

At the completion of the CONAES effort (National Research Council, 1979), the most prominent conclusions of the study were

• Energy demand elasticity is large, but only in the long term, corresponding to turnover rates of capital stock. Hence, large price changes lead to an inevitable lessening of the traditionally strong linkage between the growth rates of economic activity and energy consumption, for example, as measured by the ratio of energy to gross national product (GNP). In other words, as the relative cost of various raw materials changes, our industrial and social system adjusts by substituting among these inputs to minimize cost. When cost changes occur slowly enough, energy substitution becomes viable.

• Future growth in energy demand depends on both prices and policy, but the most likely scenarios imply that growth rates for energy consumption will be lower than in the past. Even if energy prices remain level, advancing technology will favor more efficient use of energy.

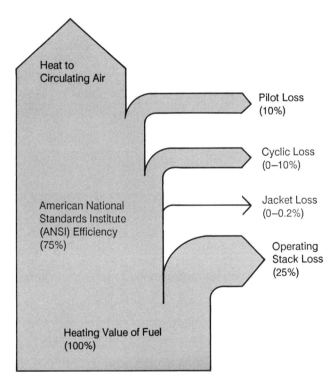

FIGURE 3 Energy flow for a gas furnace system. SOURCE: National Research Council (1979).

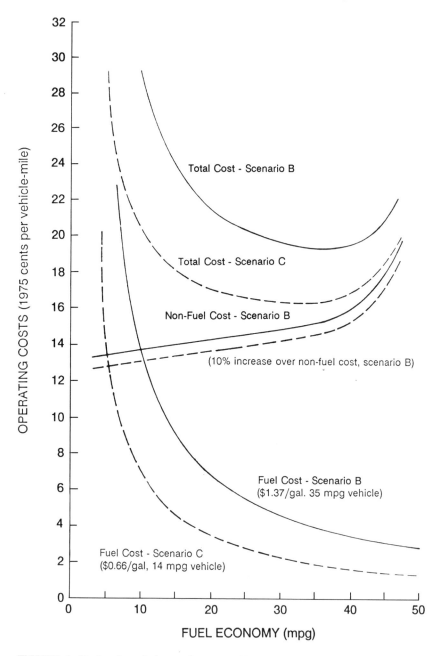

FIGURE 4 Fuel and nonfuel costs for automobiles. Marginal improvements in efficiency contribute less to total operating costs as fuel economy increases. SOURCE: National Research Council (1979).

TABLE 1 Energy Price Scenarios of the Committee on Nuclear and Alternative Energy Systems

Scenario	2010 Energy Price[a]	2010 GNP[b]	Average Annual GNP Growth Rate[c] (percent)
A	4 × 1975 price	2 × 1975 GNP	2
B	2 × 1975 price	2 × 1975 GNP	2
C	1 × 1975 price	2 × 1975 GNP	2
D	2/3 × 1975 price	2 × 1975 GNP	2
B'	2 × 1975 price	2.8 × 1975 GNP	3
A*	4 × 1975 price and progressive curtailment	2 × 1975 GNP	2

[a] Average for all the forms of energy combined.
[b] Gross national product.
[c] Average growth rates for the 1975–2010 period.

SOURCE: National Research Council (1979).

- Liquid fuels constitute by far the most important fuel resource problem.

If one looks back over the past 10 years, several reflections come to mind:

- The favorite scenario of the CONAES study (2 percent average GNP growth since 1975 and about 2 percent average increase in real energy prices) has turned out to be pretty accurate (see Table 1). Whereas the projected consumption of energy in 1986 was 76 quads (quadrillion Btu) (scenario B, Figure 5), 74 quads was actually consumed.
- The CONAES study was much too conservative with regard to the technological limits obtainable for auto efficiency (37 miles per gallon); see Figure 6.
- Industrial demand growth was greatly overestimated, perhaps because the profound changes in the structure of the economy that occurred over the 1980s were not anticipated (see Figure 7).
- Some important health and environmental issues, as they related to the efficiency of energy use, were underappreciated: for example, indoor air quality, urban and regional air quality, and the effects of chlorofluorocarbons, ozone, and other greenhouse gases.

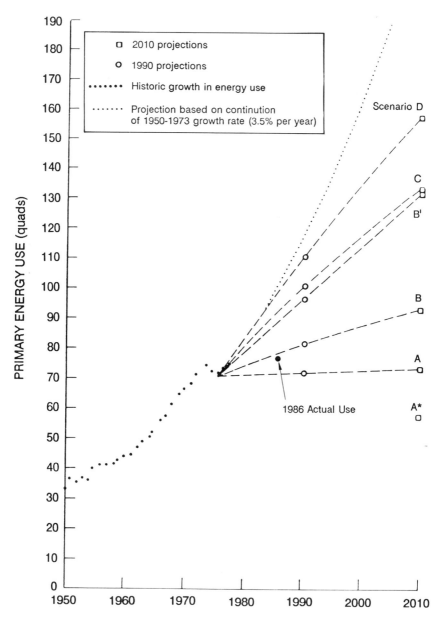

FIGURE 5 Total primary energy use projections for six scenarios. As of 1986, scenario B has proved the most accurate. SOURCE: After National Research Council (1979).

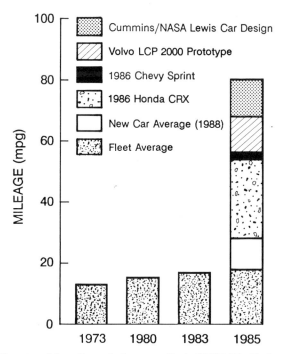

FIGURE 6 U.S. automobile mileage (miles per gallon). SOURCE: Bleviss (1988).

FUTURE IMPLICATIONS

In retrospect, investments in energy efficiency since the early 1970s have played an important role in the reduction of U.S. dependence on foreign sources of oil. Throughout the 1970s, actions taken to improve energy efficiency were often dramatic successes but occurred primarily in industry, transportation, commercial buildings, and residences when fuel was a very significant operating cost and, therefore, when the payback for such an investment could be realized very quickly. Actions were often accelerated even further by public policy incentives. Some of the actions involved changes in patterns of energy use, such as lowering thermostats, but most involved investments in technology, either retrofits of existing technology (e.g., insulating existing homes) or new investments in technology (e.g., energy-efficient new construction or automobiles with improved mileage). In many cases, more than a decade later, the returns from those investments can still be seen. Although there was usually sufficient incentive to replace capital with more energy-efficient technology on a life-cycle cost basis, the initial capital cost was often enough of a disincentive to defer the investment until the existing capital approached the end of its useful life.

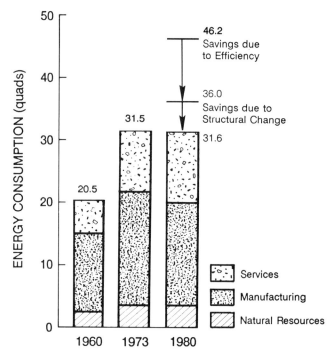

FIGURE 7 The role of energy efficiency in U.S. industrial energy consumption. In 1980, 31.6 quads was consumed, rather than 46.2, because of unanticipated savings due to efficiency gains and structural changes. SOURCE: U.S. Department of Energy (1987).

By contrast, energy efficiency gains in the 1980s, although often just as significant as those in the 1970s, have frequently been realized as incidental benefits to other investments aimed at improving the competitiveness of U.S. products in world markets. Energy efficiency investments in both the 1970s and the 1980s were, and generally continue to be, easier and more cost effective than finding new sources. Although many more opportunities for improvements in energy efficiency still exist in virtually all sectors of the economy, capitalizing on many of these opportunities may require policy intervention.

Much has changed since the 1970s. The traditional points of reference for U.S. energy security have been the two oil embargoes of the Organization of Petroleum Exporting Countries (OPEC). Those events symbolized the skyrocketing oil and gas price trends of the period. Since that time the U.S. economy has experienced a fundamental evolution of its energy characteristics. As noted, some of the changes of this period were behavioral changes in the use of energy, such as lowered thermostats, but many were more permanent structural changes, such as increases in both the efficiency and the flexibility of energy-using technologies. Partly as a result of both kinds of change, energy as a percentage of GNP has dropped more than 24 percent since 1973 (see Figure 8).

It is important to note, however, that in addition to increased efficiency of use, other forces contributed significantly to this change in energy consumption as a fraction of GNP. These forces center on the changing structure of the U.S. economy. For example, a dramatically altered trade balance during the 1980s has affected economic growth and disguised our dependence on energy (through the energy embodied in imported goods). In addition, although actions taken to improve energy efficiency in industry were important contributors to the decline in energy consumption per unit of U.S. economic output, the relative contribution of increased efficiency in industry to this decrease is not as clear because, during this same period, the U.S. economy was also (and still is) experiencing a dramatic shift in production and delivery to less energy-intensive products (see Figure 9). For example, between 1973 and 1984, imports of energy-intensive goods increased from 4 percent to more than 7 percent of total GNP (see Figure 10). Of particular note, the import of machinery grew from 15 percent of non-energy product imports in 1970 to more than 25 percent in 1985.

Similarly, changing demographics are altering the traditional energy consumption patterns of households. For example, the participation of

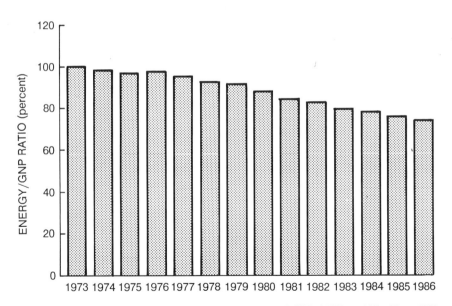

FIGURE 8 U.S. energy consumption as a percentage of GNP (1973 = 100). Since 1973, this indicator has dropped more than 24 percent in real terms. SOURCE: U.S. Department of Energy (1988).

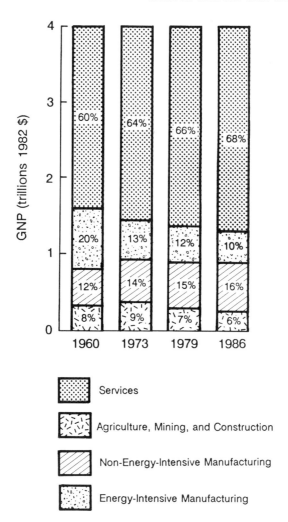

FIGURE 9 Trends in U.S. GNP (trillions of 1982 dollars), 1960–1986. Percentages of GNP indicate trends toward less energy-intensive activities. SOURCE: U.S. Bureau of the Census (1987).

women in the work force grew from 36 to 56 percent between 1976 and 1986.

Changes in the residential sector have contributed significantly to improved energy efficiency (see Figures 11 and 12). This is also true of the commercial sector. These improvements have come largely in the form of more efficient lighting and appliances, increased insulation to improve the efficiency of heating and cooling, and more intelligent management of energy use in buildings.

As noted earlier, competitive pressures on industry are encouraging investments in energy efficiency indirectly, as a consequence of efforts focused on other factors that affect overall productive efficiency. Decisions to modernize industrial plants, primarily focused on reducing labor costs, for example, are likely to trigger improvements in energy efficiency that otherwise, on their own, might not be considered cost-effective. For example, the U.S. steel industry today is very different from that of a decade ago. It has changed from high-volume production of generic steel to lower-volume production of specialized, high-value products. Hence, although the U.S. steel industry's total value of production of steel products has not declined substantially over the past decade, the composition of its output has changed considerably. On the one hand, the investment in transforming the industry has resulted in dramatically improved energy efficiency. On the other hand, the United States now imports much of its generic steel.

Transportation is the sector most vulnerable to the behavior of oil imports. Gains in the energy efficiency of transportation over the past decade have been considerable and constitute perhaps the greatest policy success in energy efficiency. Efficiency legislation passed in the 1970s

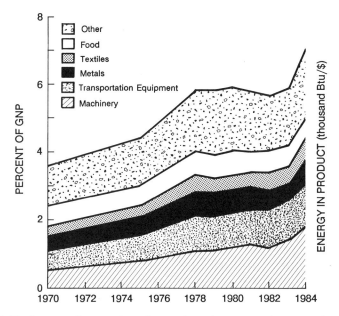

FIGURE 10 Imports of energy-intensive products (non-energy imports only; imported energy such as oil and natural gas is excluded). SOURCE: U.S. Department of Energy (1987).

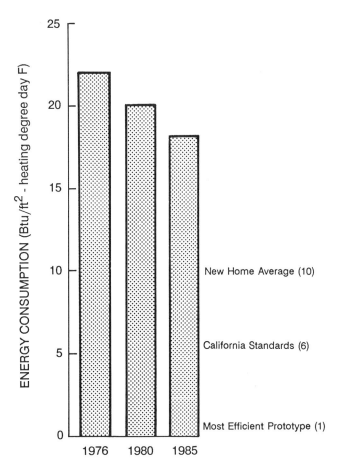

FIGURE 11 U.S. residential energy consumption (average energy to heat a single-family residence). SOURCES: Blackburn (1987) and Meier et al. (1983).

mandating corporate average fuel economy standards for automobiles has led to dramatic gains in transportation energy performance. The fuel economy of new cars today is nearly double that of the past decade. Although the replacement rate of cars has slowed, considerable future efficiency gains are also possible and perhaps essential if we wish to further reduce oil imports. For example, the fleet fuel consumption average of U.S. autos has improved steadily over the past decade to about 18 miles per gallon. With existing technology, improvements to 40 or more miles per gallon are possible, and if consumers are willing to accept smaller cars, improvements to nearly 80 miles per gallon can be achieved. Clearly there is a long way to go and much time will be required.

Finally, changes on the supply side will, of course, affect the future of demand and, hence, the prospects of future gains in energy efficiency. In particular, the traditional equation of vulnerability to a short-term disruption related to the level of oil imports has become less emphasized in policy discussions. The events of the 1970s prompted the creation of a strategic petroleum reserve (SPR) as well as industry actions to stress flexibility in energy use. For example, even industries that replaced oil use in the 1970s with the use of natural gas or other alternatives such as cogeneration often retained the capability to burn oil. Although it is difficult to quantify this fuel flexibility in U.S. industry, such a capability, along with the existence of the SPR and improved energy efficiency, all serve to lessen vulnerability to short-term oil disruption, even as increased dependence on foreign oil returns.

In the longer term, however, most trends in the economy, resource availability, and energy demand point toward dangerously increased vulnerability in the coming years. The domestic oil and gas resource base continues to decline. Moreover, the increasing concentration of world oil

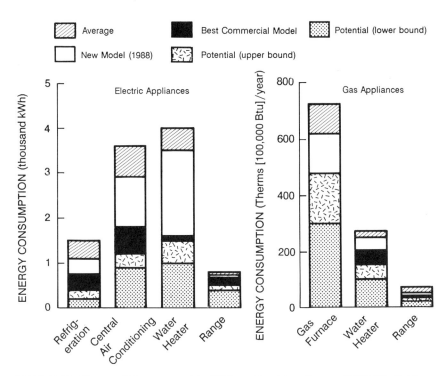

FIGURE 12 Energy efficiency potential of U.S. residential appliances. SOURCE: Goldemberg et al. (1987).

reserves in the Middle East exacerbates the problem of declining domestic reserves. Currently depressed world oil prices, as well as increased concern over nuclear power and the environmental impact of burning coal, are currently pushing industry toward a return to oil and gas. The development of new "clean-coal" technologies, smaller scale "advanced-design" nuclear power plants, and synthetic fuels has slowed considerably. Finally, lower oil and gas prices have also prompted economic and regulatory changes that could increase the use of oil and gas beyond what might have resulted from the price effect alone, for example, natural gas deregulation or repeal of the Power Plant and Industrial Fuel Use Act of 1978. In addition, lower oil and gas prices have reduced industry incentives to explore for new domestic resources, thus accelerating dependence on imports.

Ironically, because the immediate sense of urgency about energy issues has diminished considerably, some of the forces that dramatically moderated our dependence on foreign sources of fuel in the 1970s (and helped drive oil prices down) may precipitate new dependence. For example, since the easiest investments in energy efficiency have already been made, future investments may be more difficult to stimulate and perhaps require stronger policy incentives. The difficulty in capturing energy efficiency gains exacerbates long-term economic vulnerability. Nonetheless, considerable energy efficiency gains in all sectors of the economy are possible in the future and constitute the cornerstone of a comprehensive strategy for slowing the increase in oil imports in the 1990s (see Chandler et al., 1988, and Gibbons and Chandler, 1981).

CONCLUSIONS

Dependence of the U.S. economy on foreign energy sources remains the nation's most critical long-term energy problem. The economic vulnerability resulting from this dependence is partly due to poor, albeit improving, energy efficiency and a lack of flexibility in responding quickly to changing energy supply and prices. Currently the United States is in a period of relative stability for energy supplies and prices, and energy efficiency gains from new technology and new capital stock are steadily rolling in. This is a result of the economy's strong reaction to higher prices and structural shifts that led to reduced demand as well as more diversified supplies and, in turn, to surplus world production capacity for major energy supplies. Surplus energy capacity stabilized energy prices, and this stability is certainly welcome after the tumultuous economic upheaval of the 1970s.

The ability to deal with short-term disruptions made possible by the SPR, by increased oil production from non-OPEC sources, and by dual fuel capabilities in industry developed over the past decade will not affect the long-term trend toward a dramatic increase in U.S. imports of oil. Hence, the current period of stability is only a reprieve, not a justification for complacency. However, energy efficiency improvements, along with greater development of domestic oil sources, will affect this long-term trend of dependence.

In sum, the potential for improvements in energy efficiency remain dramatic, as typified by the illustrations cited earlier in heating, cooling, and lighting of residential and commercial buildings and in improved automobile efficiency. This potential will continue to expand with improvements in information technology and new materials, but significant constraints limit the rate of realizing that potential. These constraints include continued low energy prices relative to the rest of the world, long time lags in the turnover of capital to incorporate energy efficiency improvements, and falling levels of research and development initiatives in the area of energy efficiency. It could well be a time to redouble our efforts to mine the considerable resources of energy efficiency still available in virtually all sectors of the economy. A comprehensive energy policy centered on increasing energy efficiency, however, is likely to require substantial policy intervention.

REFERENCES

Blackburn, J. 1987. The Renewable Energy Alternative. Durham, N.C.: Duke University Press.

Bleviss, D. 1988. Energy Efficiency: The Key to Reducing the Vulnerability of the Nation's Transportation Sector. Washington, D.C.: International Institute for Energy Conservation.

Chandler, W., H. Geller, and M. Ledbetter. 1988. Energy Efficiency: A New Agenda. Washington, D.C.: American Council for an Energy-Efficient Economy.

Gibbons, J., and W. Chandler. 1981. Energy: The Conservation Revolution. New York: Plenum Press.

Goldemberg, J., T. Johansson, A. Reddy, and R. Williams. 1987. Energy for a Sustainable World. Washington, D.C.: World Resources Institute.

Meier, A., J. Wright, and A. H. Rosenfeld. 1983. Supplying Energy Through Greater Efficiency. Berkeley, Calif.: University of California Press.

National Research Council. 1979. Alternative Energy Demand Futures to 2010. Committee on Nuclear and Alternative Energy Systems. Washington, D.C.: National Academy of Sciences.

U.S. Bureau of the Census. 1987. Statistical Abstract of the United States: 1988, 108th ed. Washington, D.C.: U.S. Government Printing Office.

U.S. Department of Energy. 1987. Patterns of Energy Demand. Washington, D.C.: U.S. Government Printing Office.

U.S. Department of Energy, Energy Information Administration. 1988. Monthly Energy Review [DOE/EIA-0035(88/01)] (April).

Energy: Production, Consumption,
and Consequences. 1990.
Pp. 52–71. Washington, D.C.:
National Academy Press.

Implications of Continuing Electrification

CHAUNCEY STARR

National energy strategies for the 1990s should be based on an understanding of the critical role played by electricity in meeting many social objectives. This chapter describes the relationship of electrification and related technologies to some current national themes: economic growth and productive efficiency, quality of life and social change, environmental improvements, international competitiveness, and the international energy resource base. Each of these topics involves a vast complex of factors; the key role of electrification in influencing their future course and outcome will be illustrated, along with some of the implications for the biosphere, technology, societal structure, and national energy policy.

Decisions during the 1990s on national energy alternatives should be made in the context of the time scale that is characteristic of the energy field and its substructure. In the United States, the decision process to initiate the use of a new energy alternative requires about a decade, prototype construction and start-up another decade, and commercial expansion about two decades; finally, the new energy alternative may function for three decades or more. Thus, a period of roughly 50 to 70 years is characteristic of the consequences of national decisions in the energy field. Of course, less time is required for small-scale technical fixes or changes, and the time our process now takes to approve the expansion of conventional technologies (e.g., coal and hydrosystems) lies somewhere in between. The point is that a national energy strategy planned and implemented in the 1990s will only begin to have an effect a decade or more later, and will have consequences to 2050 or beyond. It is, therefore, important to project the scope of such consequences into the next century, despite the uncertainties involved. Some of the foreseeable long-range outcomes and their implications will be indicated in this chapter.

HISTORICAL PERSPECTIVE

Electricity is a unique intermediary between a variety of primary energy sources and a broad spectrum of end-use devices, many of which exist only because electricity has become widely available. The importance of electricity in the modern world is indicated by the fact that global and U.S. sales of electricity are roughly the same magnitude as the sales of oil products for end uses (globally, about $500 billion; United States, about $150 billion), even though electricity is made from primary sources of energy and therefore can be more costly than oil on an energy basis.

Electricity is used in four ways:

1. To substitute directly for primary fuel use (i.e., a Btu or joule alternative) as in space heating or electrothermal metallurgical processes;

2. To permit the use of more flexible mechanical work devices (e.g., the electric motor as a substitute for the steam engine or human labor);

3. To provide the energy necessary for dissociating molecules or exciting atoms (e.g., batteries, lasers, plasmas, electrochemical cells); and

4. To provide the key input of electrotechnical systems for communication (e.g., telephone, wireless, television), computation, and information storage and retrieval.

In a recent survey, researchers were asked to select the most significant technical advances of all time (*Research and Development*, 1987). Harnessing electricity received 37 percent of the first-place votes; antibiotics, 14 percent; and vaccines, 11 percent. Even though such perceptions are subjective, this confirms the intuitive belief of most technologists that the process of electrification during the past century has been a major factor in the sociologic and economic development of modern industrial societies. How major a factor and the implications for the future are the concerns of this chapter.

First, the basis for the connection between electrification and economic growth should be reviewed. This topic was recently studied by the Energy Engineering Board of the National Research Council (1986) through its Committee on Electricity in Economic Growth. The study confirmed that electricity is not merely an alternative energy form (i.e., a Btu option), but rather a unique and major input to productivity.

Although electricity was first used for lighting, the electrification of industry started in the early 1900s when the electric motor became a commercial product. The energy demand effect is shown in Figure 1. The

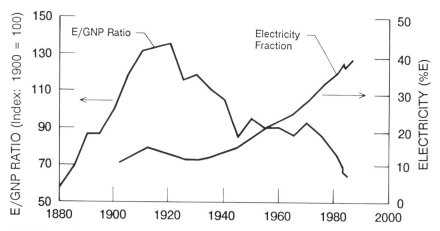

FIGURE 1 The decline in the ratio of U.S. energy consumption to gross national product (E/GNP) after 1920 is due to gains in the overall energy efficiency of the U.S. economy. At the same time, the fraction of the total U.S. energy consumption as electricity rose.

electrical fraction of the total U.S. energy consumption increased rapidly after the 1920 period. The fact that while electrification was increasing, the overall energy efficiency of the U.S. economy was also improving (as shown in Figure 1) is not often appreciated. This phenomenon seems counterintuitive at first. Electricity generation is often perceived as a wasteful process, since about two-thirds of the primary energy input is rejected as waste heat at the power plant. However, electricity-based technologies generally are highly efficient and have served to raise the net productivity from all factor inputs, including energy. The whole manufacturing system was provided the flexibility to redesign processes for optimizing production. The technological manufacturing improvements spawned by the availability of electricity helped reduce the intensity of total energy use through such productivity improvements. Therefore, the "form value" of electricity compensates for the rejected heat loss, which of course can also be used for other purposes.

If the sources of long-term growth are partitioned into two categories— (1) growth in the primary productive inputs (i.e., labor, capital plant, and equipment) and (2) growth in productive efficiency (i.e., multifactor productivity)—the latter category has accounted for more than half of the overall growth in manufacturing output since the early 1900s.

What accounts for productivity growth? Technological progress has obviously been of major importance. More specifically, the main point is that in this century technological progress has been heavily dependent on the use of electricity and electrified techniques of production.

This can be illustrated with a few statistics drawn from a major study now nearing completion at the Electric Power Research Institute (EPRI; Schurr, 1988). During the long period 1899–1985 (almost the entire twentieth century to date) when capital inputs grew at an annual average rate of 3.1 percent, electricity use grew at 8.5 percent while nonelectrical energy's growth proceeded at a comparatively slow 1.5 percent. This says that technological progress, as embodied in new plant and equipment and in overall systems of production based on such additions to capital stocks, has shown a strong affinity for energy use in the form of electricity. This preference was particularly strong during 1920–1929 when electricity grew at 10.4 percent while nonelectrical energy actually declined at a rate of 0.2 percent per year. (This was the period in which the electrification of mechanical power in manufacturing became almost total; the comparative rates reflect the fact that such electrification applied not just to new equipment, but also to the replacement of old equipment based on steam.) The detailed evidence, whether economic-statistical for all manufacturing or anecdotal for specific technologies, is quite strong that through its critical role in technological advance, electrification has been an important engine for economic growth throughout the twentieth century.

The relationship between electrification and economic growth not only is plausible from industrial case studies, but also is supported by the close correlation of electricity use with economic output measures such as the gross national product (GNP). The recent study by the National Research Council's Committee on Electricity in Economic Growth (National Research Council, 1986) has explored this subject in some depth for the United States. However, we will show that these relationships apply globally. As a guide, consider the U.S. electricity-GNP relationship summarized in Figure 2. As electrification of the U.S. economy increased over time, the slope of the electricity-GNP line became steeper. A more detailed plot of the recent period 1947–1987 is shown in Figure 3.

Obviously, it is the end use in industry that contributes to the economic output rather than electricity supply. Thus, one would expect that an increase in the efficiency of converting electricity to end uses would reduce the slope of electricity consumption versus GNP. A hypothesis to this effect, proposed in 1976, is shown with U.S. data in Figure 4. A detailed study shows that the 1973 oil embargo, which initiated fuel cost increases and thus electricity price increases, resulted in an economic drive for more efficient use of electricity, primarily in industry. The shaded trapezoid in Figure 4 shows the domain that might be achieved by improving the efficiency of electricity use. A detailed EPRI study of the technological potential for improved efficiency indicates that as much as 34 percent of U.S. use in the 1970s could be saved if all known technical means were applied (Smith, 1978). For cost reasons, roughly half of this, 17 percent, might reasonably be achieved. This 1976 chart is a convenient way to follow yearly trends.

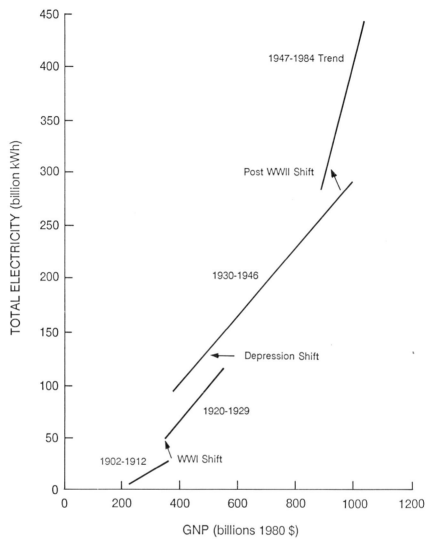

FIGURE 2 U.S. electricity generation versus GNP (lines of regression by periods, 1902–1984).

Clearly, the electricity-GNP relationship can be altered by several factors, such as efficiency, industrial mix, or commercial and residential patterns. Nevertheless, the relation provides a rough "umbrella" indicator of the technological development of any society. In Figures 5 and 6 the electricity-GNP relationship is seen to be global. Although the fine structure

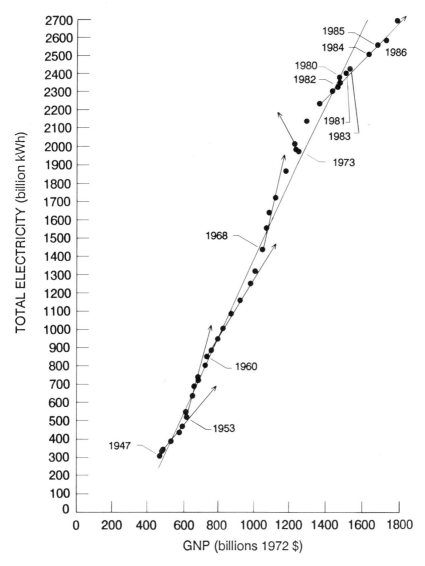

FIGURE 3 U.S. electricity generation versus GNP (1947–1987).

of this gross relationship has yet to be fully explored, there seems little doubt that electricity use and GNP are intimately related. This provides a historical basis for speculating about future trends and their significance.

A plausible future scenario requires that electrification be viewed globally, rather than limited to the United States or other industrialized

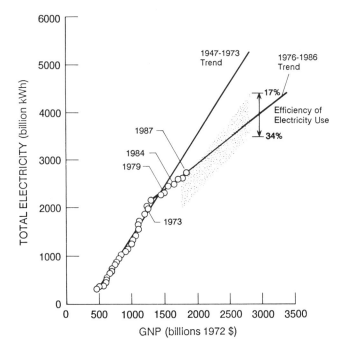

FIGURE 4 Effect of conservation on the electricity-GNP relationship in the United States. The shaded area shows the range of potential for more efficient use of electricity.

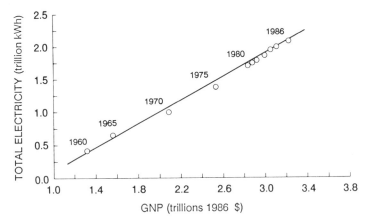

FIGURE 5 Electricity generation versus GNP in the USSR and Eastern Europe (1960–1986).

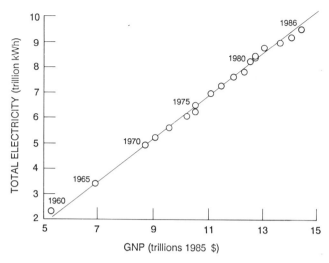

FIGURE 6 World electricity generation versus GNP (1960–1986).

countries. This global view is the important one because of the fungible nature of primary fuel resources. Further, regional aspects of electrification arise from global biospheric effects and global depletion of energy resources. Long-term energy issues should, therefore, be addressed globally.

A study by the Conservation Commission of the 1986 World Energy Conference (WEC-CC) projected the range of total energy demand to the year 2060 and the resulting fuel resource implications (Frisch, 1986). This long-term projection serves to reveal some of the basic issues associated with the characteristically slow changes in energy systems, which generally have a 50-year time constant for significant change. The WEC-CC consisted of many international expert panels in all aspects of energy systems from a variety of countries (none from the United States), supported by regional specialists. This multiyear comprehensive study focused on total energy equivalents, not electricity, but it included all the sources primary to electricity and, therefore, provided the basis for this evaluation of the implications of continuing electrification. In particular, estimates from other sources of world economic growth and population growth have been combined with the WEC-CC resource projections to describe a probable future.

It is necessary to be both skeptical and appreciative of any global projections extending about 75 years hence. Many now unforeseen geopolitical, technical, and resource changes may arise to alter radically any forecasts. Alternatively, some energy trends have had long-term stability (e.g., automobile use). Projections to 2060 may, therefore, help disclose those issues

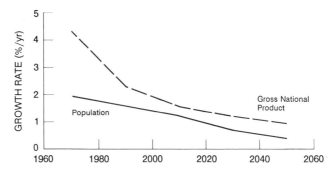

FIGURE 7 Past and projected world average annual percentage growth rates (growth rates centered in period; projections from 1985).

that justify near-term attention. Linear extrapolations of the past can also oversimplify future relationships, but in the absence of more valid models, they provide an initial framework for contemplating the future. With these caveats, the global electrification picture will now be projected.

The historical base for the projection in this chapter is the 25-year period 1960–1985. Earlier than this, world data are too uncertain. This period was used to explore three linear relationships: (1) electrical kilowatt-hours per unit of global GNP; (2) electrical kilowatt-hours per capita of global population; and (3) global GNP per calendar year. An additional projection of the world's population growth was taken from the WEC-CC study (Frisch, 1986), and it in turn was assembled from the World Bank and United Nations population forecasts. The annual growth rates of the global GNP and population provide two independent parameters for projecting global electricity. It is interesting that both of these show very modest growth rates, as illustrated in Figure 7.

It is obvious that the projection of long-term global economic growth is basic to projecting electricity consumption in the twenty-first century and thus its role in global sociologic change. To provide a general perception of the importance of economic growth, it is illuminating to consider the distribution of regional electricity use in 1985. About 31 percent of global primary energy was converted to electricity, with 33 percent efficiency in the industrial world (the Soviet Union and member nations in the Organization for Economic Cooperation and Development) and 25 percent efficiency in the developing countries. The industrial nations, with about 25 percent of the world's population, generate approximately 75 percent of the world's electricity. On a per capita basis, this means that on the average, people in the industrial world use 9 times as much electricity as residents of the developing world. The significance of these simple ratios is that future global economic growth, particularly of the developing regions, will be associated with very large increases in electricity generation.

Based on three independently derived equations relating global electricity consumption to estimates of global GNP and population growth, two independent but similar results are projected (as shown in Figure 8): by 2040–2060, electricity demand will be about 3.0 times that of 1980 (a growth rate of 1.87 percent per year during this 60-year period). The equations given in Figure 8 may be a lower boundary, because new electrotechnologies may increase the range of uses. If one assumes that the efficiency of conversion of primary fuel to electricity is not improved, this would mean that during 2040–2060, electricity generation will require as much primary energy as total world energy consumption today. Skeptical as one may be about the credibility of such long-range projections, the historical trends of the past century suggest that these may be conservative.

Even if the uncertainty of long-range projections of electricity demand is acknowledged, it is nevertheless instructive to examine their implications for principal primary fuel resources. The WEC-CC study (Frisch, 1986) examined in great detail the availability of primary fuels. Three scenarios of total energy demand were considered: high, central, and zero growth. The central scenario is considered the most likely and will be used here because it is consistent with the extrapolation of 1960–1985 trends. Its energy demand growth rate averages about 1.3 percent per year for 1980–2060, and appears both modest and reasonable compared with the 1960–1980 growth rate of 3.9 percent per year. Based on this central projection, the WEC-CC estimated the changing contributions by competing primary energy sources, giving consideration to proven, probable, and speculative supplies, as well as to the economic constraints on their production. The resulting distribution is shown in Figure 9. The message is simply that the role of coal as the world's principal fossil fuel will become more prominent in the coming decades. To quote this study (Frisch, 1986:63):

> Today, when we consider the glut of oil available and the spectacular falling price of a barrel, abundant energy might well seem to be a reality. This is true, however, only so long as we confine our perspective to the limits of this century. As soon as we pass beyond the year 2000, the appearance of reality can be seen for what it is, an illusion.
>
> The first shortages are nigh. Close at hand, the pressure on supplies grows fiercer, the struggle to find resources intensifies. The first to come under threat are the hydrocarbons and, indeed, also uranium unless breeders come to the rescue. Regionally, it is the big importers, the western countries, that face difficulties but, as time goes on, the Third World and the East will also become more and more exposed.

Actually, the Third World countries may be the most sensitive to resource scarcities, because fuel imports play a very large role in their economies.

When the WEC-CC results are applied to the electricity projections, the major fuel sources are distributed as shown in Figure 10. The assumed availability of natural gas for fuel purposes does raise a question as to whether such a high-quality resource should be consumed in this manner, instead of being kept as a raw material for the production of petrochemicals and synthetic products. Nevertheless, the important role of fossil fuels is evident. In 2060, this projection calls for 1.64 times as much annual fossil fuel consumption for electricity generation as was used in 1980. In Figure 10 it is assumed that the world's hydroelectric energy is 5 times as much and nuclear energy 18 times as much as was available in 1980. Whether hydroelectric growth will be this great, even if feasible at 2.0 percent per year, or nuclear energy will expand this much at 3.7 percent per year, depends on many existing constraints. Even in such an authoritarian and centrally planned economy as the USSR, major hydroelectric and nuclear

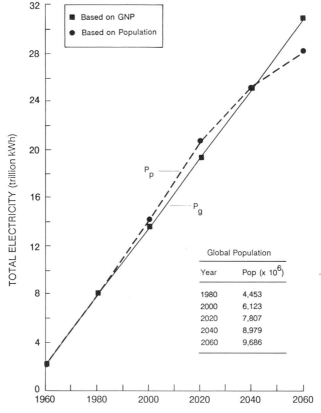

FIGURE 8 World electric power consumption—actual and projected (from 1980 to 2060), based on GNP (P_g) and population (P_p). $P_g = -2,076 + 0.8202(\text{GNP})$, where GNP in billions of 1985 U.S. dollars is given by GNP $= -679,252 + 349.218(\text{calendar year})$. $P_p = -9,466 + 3.872(\text{Pop})$, where Pop is given by the inset table.

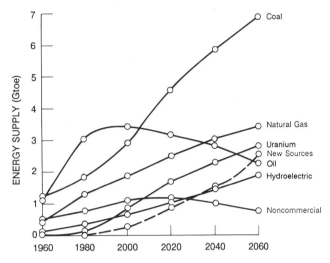

FIGURE 9 Evolution of world energy supplies (projection from 1980). SOURCE: Frisch (1986).

power plant projects have been modified because of general concern with potential long-range environmental and ecologic effects. If this becomes a global trend, then the role of fossil fuels, and coal particularly, increases substantially—regardless of whatever greenhouse effects they may produce. Although nuclear power avoids the atmospheric environmental impacts of fossil fuels, its public acceptance depends on establishing the safety of the whole nuclear fuel cycle. Perhaps the cost of electricity will eventually be the dominant determinant, driven by the economic scarcity of fossil fuels. In any event, the history of international calamities and uncertainties in long-range strategic planning would suggest that all technical options should be kept viable and that each may have a useful niche in the future global mix of fuels for electricity generation.

BIOSPHERIC IMPLICATIONS

Let us now consider the broad implications of such long-range growth in global electrification. It is evident that the most certain consequence will be a significant annual increase in fossil fuel combustion products emitted to the biosphere, with a resulting greenhouse effect. Whatever climatic change the continuing growth in some atmospheric gases (e.g., carbon dioxide, methane) is projected to produce, such changes will be accelerated by the future increase in global fossil fuel use. There is at present no economically practical technology that can be used to remove carbon dioxide as an end product of fossil fuel use. Although it is technically

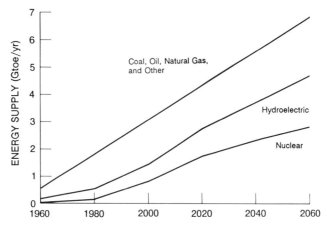

FIGURE 10 Projected mix of available world primary energy sources for electricity generation (projections from 1980).

possible to strip carbon dioxide from power plant flue gas, the ultimate conversion of the carbon dioxide to storable forms requires large energy inputs and is very costly. The permanent storage of carbon dioxide is not foreseeably resolvable. Reduction in its rate of production may result from seeking an increase in the efficiency of fossil fuel conversion, and this may delay the onset of the inevitable biospheric effects by a decade.

It is important to recognize that any reduction in electrification growth by using fossil fuels directly for end purposes (i.e., a low-electrified economy) is apt to increase, rather than decrease, the total emission of pollutants. Except for the use of passive solar heating where feasible, the system efficiency of primary resources conversion to end functions is higher for electrified systems because of the relatively high efficiency of electrical devices. Also, electrification permits the use of nonfossil resources such as nuclear, hydroelectric, and solar power for electricity production. Further, centralized generating plants based on fossil fuels usually permit a much higher quality of pollution control than is feasible with small dispersed activities. The goal, then, is to minimize pollution by promoting the most efficient generation and use of electricity.

On the brighter side, technology is available to reduce the output of other noxious atmospheric effluents such as the oxides of sulfur and nitrogen that produce acid rain. In the coming decades, the new-technology fossil fuel plants should reduce these effluents to minimal levels. Coal ash will still need disposal, but as new uses are found for this material, it can become a resource material.

A further uncertainty in long-range electricity projections arises from the effect of projected climatic change by altering present electric load patterns. A study now under way for EPRI indicates that this impact on electric utilities could be significant and that these changes could be evident within several decades. The net effect is to shift the structure of electrical demand in such uses as agricultural irrigation pumping, air conditioning with heat pumps, and other weather sensitive loads. If climatic changes also cause demographic movements of industry and populations, a corresponding shift of transmission networks will be needed as well.

A major technical factor affecting large increases in regional electrification is the increasing scarcity of inland cooling water to accept the by-product heat from either fossil- or nuclear-fueled power plants. Regional constraints on water use and ecologically acceptable water temperature increases have already limited power plant expansion plans in many areas. This has created continuing engineering interest in the development of dry cooling towers (atmospheric heat rejection), and a parallel interest in raising the Carnot efficiency of the thermodynamic cycle. This would also have the salutary effect of reducing the total effluent outputs per unit of electricity. Major progress toward this objective depends on the development of very high temperature materials, such as workable ceramics, which represent one of the active frontiers of materials research. More distant is the use of high-temperature thermionic converters for additional efficiency gains.

Another technical option for mitigating some of these undesirable constraints is the current development of direct conversion to electricity by the electrochemical fuel cell (U.S. Department of Energy, 1986), which avoids the Carnot cycle limits and may achieve efficiencies of approximately 60–80 percent. Materials lifetime and economics are current issues, but another decade of development may produce competitive commercial units. The fuel cell needs hydrogen, foreseeably produced from hydrocarbon fuels, and an oxidant such as the oxygen in air. Three approaches are now being developed: the low-temperature phosphoric acid, and the high-temperature molten carbonate and solid oxide fuel cells.

Gas turbines work well with carbon monoxide. This, fortunately, befits the output of a coal gasifier, which can use coal and water to produce equal amounts of hydrogen and carbon monoxide. Thus, given an economically functional phosphoric acid fuel cell, the efficient way to use coal is through gasification, with carbon monoxide fueling a thermodynamic combined cycle generator (gas turbine and steam cycle), and with hydrogen going to the fuel cell. This combination might provide a 25 percent efficiency improvement in the use of coal for electricity generation. More options become possible because of the ability of high-temperature molten carbonate and solid oxide fuel cells to use carbon monoxide (or methane) directly with the added benefit that by electrode surface reactions with water, hydrogen is produced.

TECHNOLOGY IMPLICATIONS

All of us are familiar with the role of electrification in a modern society. Personal productivity is generally enhanced by a variety of electrical devices. The more recent applications in industry are described in a series of EPRI reports (EPRI, 1984–1987). Of special interest is the potential of the more recent developments such as lasers, plasma torches, superconductors, and new materials.

Lasers provide a highly collimated energy beam that can be tuned over a wide range of frequencies ranging from infrared to ultraviolet. They can be used to stimulate chemical reactions and separate isotopes by selective photoexcitation, photoionization, or photodissociation of atomic or molecular vapor. When used with computerized controls, lasers are replacing conventional cutting tools with commercial success.

The plasma torch can reach a temperature of almost 8000°F (4400°C), about twice the temperature of a flame. Plasma torches could dissociate the molecular structure of most compounds. A plasma-fired cupola for iron foundry melting is one of the many new and potentially large present applications. The use of the plasma torch to destroy toxic wastes has been demonstrated to be extremely effective, and this "pyroplasma" technology will eventually become a major tool for environmental waste control. The laser is scale limited at present and thus not adapted for a large throughput of material. However, the application potentials are sufficiently attractive that low-cost modular units may eventually be developed.

More speculative is the possibility of future commercialization of electric laboratory processes such as the synthesis of new products by using electrochemistry and organic plasma chemistry (Schurr, 1983). An electrode reaction or a controlled plasma can provide the energy needed for molecular excitation and bond breaking, with subsequent stabilization of resulting species to provide new products. Electrochemical and plasma processes have the ability to produce high-energy electron level excitation, thereby creating new species and reactions. Great progress has been made in the laboratory synthesis of new products, but so far the energy efficiency and rate of production are low. The development of large-scale electrochemical equipment is the challenge for the twenty-first century.

A recent electrochemical development appears to have promise for removing toxic chlorinated chemicals from organic waste. By electrochemical stripping of the chlorine atom from chloro-organic compounds, their inherent toxicity is destroyed. A commercial prototype dechlorinating plant has been built to treat pesticides.

Perhaps the most exciting near-term electrotechnical development is the electric automobile. Electric vehicles have been used since the earliest days of the automobile, but their use has been severely limited by the electricity storage capacity and lifetime of the conventional battery. In the past decade, intensive development of the lead-acid battery and the electric vehicle has made the combination marginally competitive for short-range urban uses and prototype demonstrations are now under way. Improvements in the traditional nickel-iron cell have succeeded in producing a battery with 50 percent more energy storage per unit weight than the lead-acid cell. It is expected that within a few years, a demonstration of a nickel-iron powered commercial van with a possible 100-mile range will be made. It appears very likely that urban use of the electric automobile will be common by the turn of the century.

Another long-range technical challenge is the fixation of nitrogen by electrical means. Low-cost fertilizer is one of the key requirements for meeting the food demands of the inevitably increasing global population. Historically, primitive arc devices were used in the United States and Sweden to fix nitrogen from the atmosphere. This was superseded by the much more economical synthetic ammonia technology. Perhaps the ongoing development of lasers and plasmas may lead to commercial nitrogen fixation which, with phosphate rock, could provide the nutrients for a future global food supply.

SOCIOLOGICAL IMPLICATIONS

The electrification of industrial societies has resulted in profound changes in their social structure. Most dramatic among these has been our emancipation from the solar day—its effect on our living patterns is obvious. The revolutionary change in manufacturing produced by the advent of the electric motor has already been described. Electrically powered small machine tools also made possible the decentralization of small-scale specialty manufacturing.

Beyond manufacturing, however, the same electric motor development provided the mechanical power for vapor-compression refrigeration units for commercial and residential use. The consequent sociological restructuring has been massive. The effect of refrigeration on the processing and distribution of food is clearly evident; it has radically increased our available sources of nutrition as well as their economics and healthfulness. Refrigeration has also totally altered our food supply and delivery systems.

Air cooling has completely changed the demography of the southern portion of the United States, as well as becoming an integral part of life-styles in hot weather (e.g., air-conditioned cars, offices, restaurants, schools, theaters, and homes). Electricity-powered cooling systems have expanded the availability of friendly living space in otherwise inhospitable climates. The continuing development of the electric reversible heat pump now provides a single device that can both cool and heat, and can do so very efficiently, so that completely electrical, space temperature conditioning is now becoming more commonplace and economical.

The electrification of communications and information systems has steadily expanded the individual's span of accessible knowledge and recreation.

Each of these areas of electrification deserves much study to assess thoroughly its sociological impacts. In this limited discussion, it is sufficient to make the general point that electrification in the twentieth century has profoundly changed our society. The interesting speculation is the degree to which the twenty-first century will see continued societal changes from this electricity-based expansion of accessible food, space, time, knowledge, and productivity.

In looking forward to the twenty-first century, the most significant electrification outcome is likely to be its contribution to global economic growth and increased per capita income. This could be the strongest force for improved global health, education, and public welfare. Such broad benefits are not usually perceived as causally related to electrification, but our studies indicate the relationship is both real and powerful, given a societal environment conducive to economic and industrial growth.

One can also predict substantial mitigation of many of the current environmental problems of industrial societies by the application of foreseeable electrotechnologies. The most obvious is the revival of urban electrical transportation systems as a means of reducing air pollution and traffic congestion. The electrical passenger train, both surface and subterranean, has again become the technically desirable transportation mode for connecting the inner city with the suburbs, and this will become more commonplace as the traffic density on surface roads increases. Further environmental benefit can be gained by using the by-product heat from electricity generation for urban district heating or other local uses.

In the city, the air pollution created by the gasoline-fueled automobile has become an incipient health hazard. The electric automobile is now a developmental reality. As previously mentioned, the commercial availability of improved storage batteries (e.g., nickel-iron) could provide a major alternative to the gasoline-fueled automobile for urban use. A study of the potential role of a successful electric automobile (100-mile range) indicates that it could satisfy 92 to 96 percent of the average family's trips and could cover approximately 66 to 74 percent of the miles traveled annually (Horowitz, 1987). Such a major transportation transition, encouraged for urban pollution reduction, could substantially raise electricity demand and reduce national oil needs by as much as one-half. This is a current technological development that could significantly change long-range electricity roles and the geopolitics of oil.

Electrification of communications will facilitate urban decentralization. It is already evident that advanced communications techniques permit any individual to function intellectually from any location, through communications links with other individuals, libraries, data banks, universities, service centers, markets, financial services, entertainment, and so on. Eventually, this could mitigate urban population density and its attendant environmental stresses. Two-way television will become commonplace as the full capacity of fiber-optic networks is achieved. The terminal equipment is costly today, but rapid progress can be anticipated to lower its cost substantially. Thus, video conferences, shopping, education, and other activities may be conducted without personal travel. Low-cost and versatile communications may also help to strengthen the family ties of members separated by a long distance.

There will, of course, continue to be many activities that require the assembly of large groups in urban centers. Transportation costs for raw materials, finished goods, and services play an important role in determining the need for centralized production and management. Thus, the pressures for urban decentralization will be balanced by counterpressures for the maintenance and revitalization of urban centers. Electrified "people movers" such as trains, subways, elevators, escalators, and walkways facilitate more efficient infrastructure for high-population-density urban communities.

These few examples of the possible future implications of electrification have focused on their potential for societal change. The rapidly developing information and electrification technologies have combined to produce a foreseeable spectrum of future technical opportunities for achieving global societal goals. In an unpublished paper entitled "Future Imperfect," Robert M. White (1986), president of the National Academy of Engineering, perceptively discusses the present era of technological change and its interaction with economic, political, and social change. In that paper, he suggests that we may not fully perceive that we are already in a period of technological innovations so sweeping that they will "transform the institutions of society, the general welfare of people, the economies of nations, and the fate of individual industries." Electrification is by no means the chief element of such a technological revolution. However, in combination with the parallel development of other technologies, it is clearly a crucial component in the creation of future opportunities to effect changes in our global societies.

IMPLICATIONS FOR NATIONAL ENERGY POLICY

Electrification is so closely interwoven with other energy systems that its future course will be influenced by the directions taken by all other energy programs. The global issues discussed in this chapter, which are pertinent to electrification, are similar to the issues for a national energy policy. These may be summarized in terms of two strategies:

1. Societal strategy—establish long-range objectives and priorities:

- economic growth;
- environmental improvement: both regional and global;
- geopolitical balance of international versus national markets;
- transition methodology for government and industry; and
- time scale for objectives.

2. Technological strategy—develop energy supply and demand alternatives for adaptation to future constraints:

- electrification a dominant mode, except for passive solar heating;
- stimulation of nonfossil resources such as hydroelectric, nuclear, and solar energy;
- risk-taking incentives for research and development; and
- program stability in government and industry.

As previously discussed, the long-time characteristics of energy systems require very long-range strategies for achieving both societal and technical objectives. The duration of such strategies must be very much greater than the short-term, two-, four-, and six-year election periods of our political processes. We need a national energy policy whose directions can be maintained and supported for decades by both government and industry.

As a more general comment, it should be emphasized that although applied science and engineering can provide the technical tools, their use depends on the initiatives and support of industrial, political, and social institutions. Familiar as this thought may be, one must continually emphasize the need for institutional flexibility to achieve the benefits made possible by the opening of new frontiers through technological progress. This requires a long-term societal commitment to achieving common goals. A historical review of the relationship of energy use and social development throughout human civilization (Starr, 1971, 1979) illustrates that energy has always served as a scaffolding for economic progress. Unfortunately, during the past two decades, the historical recognition of energy as a positive good has undergone a reversal. Once again, the productive contribution of energy—used efficiently, of course—must be recognized. It is in this context that electrification appears to be the most effective mode for the use of future energy resources.

REFERENCES

Electric Power Research Institute. 1984. Microwave Power in Industry. EPRI report no. EM-3645. Palo Alto, Calif.: EPRI.
Electric Power Research Institute. 1985. Electricity and Industrial Productivity: A Technical and Economic Perspective. EPRI report no. EM-3640. Palo Alto, Calif.: EPRI.
Electric Power Research Institute. 1985. Induction Heating of Metals: State-of-the-Art Assessment. EPRI report no. EM-4131. Palo Alto, Calif.: EPRI.
Electric Power Research Institute. 1985. Resistance Heating of Metals: State-of-the-Art Assessment. EPRI report no. EM-4130. Palo Alto, Calif.: EPRI.
Electric Power Research Institute. 1985. Vacuum Melting of Metals: State-of-the-Art Assessment. EPRI report no. EM-4132. Palo Alto, Calif.: EPRI.
Electric Power Research Institute. 1986. Electroforming of Metals: State-of-the-Art Assessment. EPRI report no. EM-4568. Palo Alto, Calif.: EPRI.
Electric Power Research Institute. 1986. Electron Beam Processing of Metals: State-of-the-Art Assessment. EPRI report no. EM-4526. Palo Alto, Calif.: EPRI.
Electric Power Research Institute. 1986. Plating, Finishing, and Coating: State-of-the-Art Assessment. EPRI report no. EM-4569. Palo Alto, Calif.: EPRI.
Electric Power Research Institute. 1986. Radiation Curing: State-of-the-Art Assessment. EPRI report no. EM-4570. Palo Alto, Calif.: EPRI.
Electric Power Research Institute. 1986. Resistance Heating of Nonmetals: State-of-the-Art Assessment. EPRI report no. EM-4915. Palo Alto, Calif.: EPRI.
Electric Power Research Institute. 1987. Radio Frequency Dielectric Heating in Industry. EPRI report no. EM-4949. Palo Alto, Calif.: EPRI.
Frisch, J. R. 1986. Future Stresses for Energy Resources. World Energy Conference—Conservation Commission. London: Graham & Trotman.
Horowitz, A. D. 1987. Exploring potential electric vehicle utilization: A computer simulation. Transportation Research 21A(1):17–26.
National Research Council. 1986. Electricity in Economic Growth. Energy Engineering Board, Commission on Engineering and Technical Systems. Washington, D.C.: National Academy Press.
Research and Development. 1987. December:65.
Schurr, H. 1983. Application of nonequilibrium plasmas in organic chemistry. Pp. 1–51 in Plasma Chemistry and Plasma Processing, Vol. 3, No. 1. New York: Plenum.
Schurr, S. 1988. Electricity in the American Economy: An Agent of Technological Progress. Palo Alto, Calif.: Electric Power Research Institute.
Smith, C. B., ed. 1978. Efficient Electricity Use. Elmsford, N.Y.: Pergamon Press.
Starr, C. 1971. Energy and power. Scientific American (September).
Starr, C. 1979. Collected Readings in Energy. San Francisco: W. H. Freeman.
U.S. Department of Energy. 1986. Fuel Cells Technology Status Report. DOE/METC-87/0257; (DE 87006525). Washington, D.C.: U.S. Department of Energy.
White, R. M. 1986. Future Imperfect. Paper presented to the Electric Power Research Institute, Monterey, Calif., August 19, 1986.

2
Environmental Issues

Global Environmental Forces

THOMAS C. SCHELLING

Greenhouse warming is global in at least two respects. First, carbon dioxide (CO_2) and the other gases released or withheld anywhere on earth disperse rapidly into the global inventory. The location of origin makes no difference. Second, the effect will be a change in global circulation of air and water. Although the mean rise in atmospheric temperature is commonly used as an index of climate change, the change in temperature differential between equatorial and polar regions may be a better measure of global environmental forces.

The standard point estimate of global warming for a doubling of the concentration of CO_2 in the atmosphere is 3°C (National Research Council, 1982). But it is usually estimated that the warming in the polar regions associated with this 3-degree average change might be 8 or 10 degrees, whereas the change in atmospheric temperature near the equator might be closer to 1 degree (National Research Council, 1982). Offhand this sounds like a welcome dispersion of temperature change: it will mainly get warmer where it is already very cold and warm up the least where it is already hot. But more significant is that it is the temperature gradients between equatorial and polar regions that drive the winds, which in turn drive the oceans, and a change of 7 or 8 degrees in the mean temperature *difference* will change the atmospheric and oceanic circulation much more than would a uniform global rise in atmospheric temperature. Most climates may get warmer, some will undoubtedly become cooler. But the observed changes will include not only temperature and temperature variation from season to season and year to year but also, probably more importantly, the amounts, the seasonal distribution, and the year-to-year variation in rainfall, snow, wind, fog, sunlight, humidity, and storms.

For the purpose of comparing forthcoming changes in climate with changes experienced in the past, the mean global atmospheric temperature is probably not only a reliable index but something of a measure of magnitude. Using the commonly accepted 3-degree rise from a doubling of the atmospheric concentration as an approximation to what may be forthcoming, the ensuing temperature will not only be well outside the range of atmospheric temperatures experienced in the past 10,000 years but may be several times the range of temperature variation experienced in that time. This observation is frequently expressed, and correctly, as a change in climate greater than any that mankind has experienced since the dawn of history. It is expressed more accurately as changes in *climates*—plural not singular—because different climates around the globe will change differently.

Without belittling the unprecedented nature of such climate changes or the prospect of some change that is not gradual but catastrophic, it is fair to point out that most people will not undergo in the next 100 years changes in their local climates more drastic than the changes in climate that people have undergone during the past 100 years. No climate changes are forecast that compare with moving from Boston to Irvine, California, or even perhaps from Irvine to Los Angeles. The Goths and the Vandals, the Romans and the Vikings, the Tartars and the Huns migrated through more drastic changes than any currently anticipated; Europeans who migrated to North and South America similarly underwent drastic climate changes. In this country in 1860 barely 2 percent of the population lived outside the humid continental or subtropical climates; in 1980 the percentages outside these zones had increased from 2 percent to 22 percent.

Furthermore the microclimates of urbanized Tokyo, Mexico City, and Los Angeles have not deterred their population growth; the microclimates of London and Pittsburgh changed dramatically during the century before 1950 and have changed again almost as dramatically since then. Even urbanization itself, without the associated air pollution, changes the conditions created by climate. Most Americans, Europeans, and Japanese never experience muddy roads anymore.

The expectation is that climates will change gradually, both over time and over space (National Research Council, 1983:Ch. 1–3). The climate of Nebraska may gradually change into the current climate of Kansas, not into the climate of Massachusetts or Oregon. Climates will "migrate." This expectation is on the whole reassuring, but it could be mistaken. The models used in the computer simulation of climate may be incapable of producing discontinuities because the current state of meteorological knowledge is confined to continuous processes. There may be no reason to expect discontinuities, but the fact that the models produce no discontinuities may reflect an inability to design models based on the state of the art that can discover such phenomena.

Aside from a possible rise in ocean level, which I shall discuss presently, the most predictable physical and economic consequences of climate change will be in agriculture. By "predictable" I mean not that the actual changes can be predicted but that it can be reliably predicted that there will be changes. These will be changes in rainfall, winter snow for summer irrigation, humidity, daylight and cloud cover, and perhaps the health and comfort of livestock.

There is no reason to believe that the revolutionary improvements in agricultural productivity that have developed over the past 75 years and that in many cases have spread worldwide will not continue. Depletion of soils may continue, but control over plant and animal genetics and the possible production of new proteins may drastically change for the better what crops people will grow and what foods they will eat 50 or 100 years from now. An increase in the cost of food production by 5 or 10 percent, even 20 percent, which would be a somewhat extravagant estimate, may easily be offset many times over by another century's improvements in agricultural productivity.

There will undoubtedly continue to be parts of the world that are intractably poor and dependent for a livelihood largely on local production of food or other climatically dependent crops. These countries may have little of the capacity to adapt that the more advanced countries can afford. So even if the damage to food production may not average enough on a global scale to be cause for alarm—may not even be noticeable—there may be particular areas in which the damage to agriculture coupled with population growth could severely retard progress. (Population growth may be the more serious.) This situation may demand foreign aid to the poorest countries. I would neither expect nor recommend foreign aid directly related to hardships induced by climate change, but rather aid to the poorest.

What I have said so far will sound to many readers as insufficiently alarmist. "Optimistic" it may appear. One reason for the unexcited tone, which I shall elaborate shortly, is pessimism, not optimism. I do not believe that serious measures will be taken over the next quarter century to curtail the emissions of carbon into the atmosphere. I do not believe that even an alarmist appraisal will lead to a substantial policy response. I therefore do not believe that exaggerating the dangers will serve a useful purpose.

But there is, I acknowledge, another reason why my assessment is so mild. As I mentioned earlier, I am attempting to assess predicted changes, and it may be that our climate models predict only what we understand well enough to include in the models. Maybe we are also good at adapting to phenomena we understand, as well as good at predicting them; and the ones we do not understand well enough to predict will cause difficulty because we do not understand them well enough to adapt. In other words, there is bias in our assessment of dangers: those we understand well enough to perceive we understand well enough to overcome, those that we have no hints of may be the dangers we would least know how to meet and overcome. Reduced rainfall in Kansas 25 or 50 years from now we may adapt to with moisture-conserving agricultural techniques, genetically altered crops that require less moisture, or the acquisition and transport of water. The phenomenon is familiar, the adaptations are familiar, and the predictions are based on familiar principles of meteorology. The "collapse" of the West Antarctic Ice Sheet would be an altogether different phenomenon.

As recently as 15 or 20 years ago, the accepted estimates were that the grounded ice—ice resting on the sea bottom and rising a kilometer or more above sea level—might, with a warming of the oceans attendant upon a warming of the atmosphere, slide or glaciate into the ocean within 75 years, causing a 20-foot rise in sea level. Like seismology in response to the test-ban controversy of the 1950s, glaciology has advanced in the past decade or two, assisted by satellite sensing, and the currently accepted estimates are that if that grounded ice should be added to the ocean level it is likely to be gradual and to take several hundred years. The urgency of that particular danger is thus reduced by an order of magnitude (unless further rapid advances in the relevant glaciology bring comparable changes in estimates back in the opposite direction). What is worrisome is that there may be other phenomena, perhaps, like the ocean level, not being perceived as "climatic," that could be as devastating as a 20-foot rise in sea level and that will not, upon further inspection, yield to more benign estimates.

When asked for an example, I can of course protect myself by pointing out that predicting the unpredictable, foreseeing the unforeseen, especially as an amateur, cannot be demanded of me. But when I am in a mood to worry I think about possible changes in the Gulf Stream and the Japanese current. The current global circulation models, as I understand it, do not include changes in the direction and velocity of ocean currents, and I am not sure that enough is known about the response of ocean currents to changes in wind patterns to predict whether there may be *catastrophes*, that is, flipflops from one equilibrium to another, rather than gradual change. Thus, there may be a missing feedback loop from warming to winds to currents to climate that, when added to the current models, will produce something more worrisome than the migration of the climate of Kansas to South Dakota.

As I said at the outset, the problem is global; and that is why it is exceedingly unlikely that anything substantial will be done to curtail fossil fuel emissions. Any nation that attempts to mitigate changes in climate through a unilateral program of energy conservation or fuel switching (or expensively scrubbing CO_2 from smokestacks) in the absence of some international rationing or compensation arrangement, pays alone the cost of its program while sharing the benefits with the rest of the world. Consider the Federal Republic of Germany, which accounts for about 4 percent of world's energy consumption and just about 4 percent of each of the three fossil fuels, coal, oil, and natural gas. If that country took the drastic step of reducing by one-third its consumption of fossil fuels, the cost in lost productivity and consumer welfare, even if it were done gradually over a period of two decades, could be equivalent to 3–4 percent of its gross national product while the concentration of CO_2 in the atmosphere would be reduced by barely 1 percent. Even for the United States, the largest energy consumer of all, phasing in a one-third cutback in fossil fuel consumption over the next 20 years at a cost perhaps equivalent to $150 billion or $200 billion per year at today's prices and income levels, would reduce emissions worldwide by less than 10 percent. The time to a doubling of CO_2 in the atmosphere might be reduced from something like 85 years to 80 years. I think it is a fair estimate that for no *individual* country, with the arguable exception of the United States, is it economical to curtail CO_2 emissions unilaterally in the interest of retarding climate change.

Any significant effort to curtail emissions would require an international rationing regime, covering the larger fraction of world energy consumption, to ration the consumption of energy, or the consumption of fossil fuels, or the consumption of carbon, in some manner that could confidently be expected to remain in force long enough to be effective, say 50 years or more. It would have to include the Soviet Union, it would have to include the People's Republic of China, and it may well have to include the Organization of Petroleum Exporting Countries (OPEC). It would require mandating compliance on the part of scores of nations that would greatly prefer to be outside the regime. And it would require for many nations trading urgently needed economic growth now for the dubious future benefits of a rationing scheme that depended on a more disparate membership than even that of OPEC. Eventually, because most of the world's known coal resources are in the Soviet Union, China, and the United States, the scheme would require those three nations to collaborate effectively and indefinitely as a cartel.

The political likelihood of solid and confidently expected collaboration of that kind would be approximately zero if energy were a homogeneous commodity consumed uniformly worldwide. But to put in effect a rationing scheme the impact of which will begin to hurt and be effective only after several decades of energy growth would require dealing with economic growth itself, and that in turn requires attention to things like population growth (Ausubel and Nordhaus, 1983; Nordhaus and Yohe, 1983). Do the Chinese claim that a policy of zero population growth is more than sufficient as a curtailment of energy use and that their country should therefore be exempt? Do the countries in the Organization for Economic Cooperation and Development participate as a unit, negotiating long-term shares in energy growth? Is there any chance they could be more successful than they have been with defense budgets, oil imports, or agricultural trade?

My pessimistic conclusion is that nothing of the sort is going to happen. I do not believe the Montreal Protocol on Substances that Deplete the Ozone Layer, signed in September 1987, is any harbinger for suppression of CO_2. Economically what is at stake is two or three orders of magnitude greater for fossil fuels than for chlorofluorocarbons (CFCs) and the prospects for technological replacement of CFCs are much brighter. (The Ozone Protocol does illustrate the need for worldwide collaboration to make restrictions worthwhile: the treaty takes effect only when ratified by nations representing two-thirds of world consumption.)

If world politics change as much in the next 75 years as in the past 75, a global fuel regime of some kind may become possible, but none is now foreseeable. If I am wrong, and world rationing of fossil fuels becomes economically and politically feasible, we shall still face the prospects for climate change. There is absolutely no possibility that fossil fuel emissions can cease altogether in the foreseeable future, and even the most optimistic could hardly hope that fuel emissions would stop growing within the foreseeable future. A most ambitious goal might be to reduce by half the growth rate in fossil fuel emissions. (As the fraction of fossil fuels represented by petroleum and natural gas declines over the coming century, fossil fuel consumption will have to increase at less than half the unrestricted growth rate in order that carbon emissions be only half what they might otherwise be.) A not unreasonable estimate, for purposes of illustration only, of growth in fossil fuel consumption over the next half century might be 2 percent per year, a rate at which the atmospheric concentration of CO_2 might double in about 85 years, reaching 50 percent elevation in about 50 years. Holding emissions to 1 percent growth would carry us beyond the middle of the next century before we reached concentrations half again as great as today's. The implied curtailment in emissions, at 1 percent compared with 2 percent, would be 10 percent at the end of the first decade, 25 percent at the end of three decades, and 40 percent by the end of five decades. That seems to me to be the outside limit to what might be economically acceptable worldwide. (How that 40 percent aggregate curtailment would be shared among consuming nations I hesitate even to conjecture.)

National programs to phase in nuclear power to replace fossil fuels for electricity, even for the production of hydrogen fuels, may again become popular. But it is still hard to measure the half-life of anxiety resulting from the accidents at Three Mile Island and Chernobyl. Any new reactors will have to be economical as well as clean. Cutting the growth of emissions from 2 percent to 1 percent may well require all electric power capacity in the future to be nuclear.

Energy conservation measures deserve emphatic attention, but investments in conservation will mainly be limited to what the private economy finds economical. National or international policy will probably be limited to research, development, demonstration, and technology transmission. Energy-efficient investments may yet get a boost from another doubling or more of the price of crude oil, but that is probably not a boost to be hoped for.

What else may be done to cope with the greenhouse problem? CO_2 can be removed from the atmosphere by increasing the mass of living vegetation or by "refossilizing" timber, burying it underground or in the ocean or coating it so that it cannot oxidize. And CO_2 can be scrubbed from smokestacks at very substantial expense. Probably at enormous expense, some attenuation could be achieved in this fashion. (Some small increase in the carbon density of forests may result naturally from the enhancement of CO_2 in the atmosphere.) The concentration of CO_2 will therefore certainly increase, and at an increasing rate, and I consider it unlikely that we shall be rescued much before the concentration has nearly doubled.

The main response will be adaptation, and most of that by ordinary people and businesses. Some of the adaptation will be by governments, but local and regional governments as much as national governments. There will be changing climates to cope with, changing urbanization, changing population densities, and in most countries probably drastic changes in the ways that people live and work and transport themselves, perhaps significant changes in what they eat. Much of the adaptation will seem generally "environmental" rather than specifically climate oriented. And, of course, there is continuous adaptation to climate even when it is not changing: we change the technology and the efficacy with which we heat ourselves and cool ourselves and clean our air and protect ourselves from storms and cope with droughts and floods and dispose of snow. The pace of change may be such that people will find themselves adapting to *climate* rather than to *changing climate*. Just as businesses shift to take advantage of better productive climates, they will keep shifting to better climates with perhaps small regard for the prospects of changing climates in given locations.

There remains to be discussed a response to climate change that receives so little attention that it deserves emphasis here—direct intervention in weather and climate. When Thomas F. Malone was chairman of the Committee on Atmospheric Sciences of the National Research Council, he wrote, 20 years ago, "The possibility that large effects may be produced from relatively modest but highly selective human interventions opens up the possibility that weather and climate modification may some day be operationally feasible" (Malone, 1968:1136). And of the modification of hurricanes he said, "If five years are allowed for the development of an adequate mathematical model, five more years for assessing the consequences of interventions of various kinds, and then ten years of field experimentation for validation, it seems unreasonable to expect much before 1990, with the probabilities fair to good that a proven technology will exist by the year 2000." He added, "The probability of success in broad climate modification is likely to exceed 50 percent by the year 2018" (1968:1138), that being the 50-year mark from the time he wrote.

Most experiments with weather modification or with changing geographical features that may lead to climate change have been local and regional. That has been true of cloud seeding and would be true of the manipulation of hurricanes. In a discussion of greenhouse warming, the possibility of global intervention has to be considered. An important kind of human intervention in global climate may be efforts to change the radiation balance itself. We know it can be done: we are doing it. That is what the greenhouse discussion is all about. The fact that we are doing it unintentionally, and the fact that the consequences may not be welcome, do not contradict that we know how, at some expense if necessary, to change the world's climate more than it has changed in the last 10,000 years.

Warming the atmosphere currently is more economical than cooling it because it happens as a by-product of energy consumption that would be costly to reduce or terminate. If we were faced with a "little Ice Age" over the next century, we might be glad to get some of that CO_2 in the atmosphere at no cost and without having to negotiate climate change diplomatically.

But we know that, in principle, cooling could be arranged. Volcanic eruptions have done it. Discussions of "nuclear winter" took seriously the possibility that human activity might lower global temperatures cataclysmically. Considering the development of nuclear energy in both its explosive and its controlled uses and the feat of landing a team on the moon and returning it safely, and that we now know how to warm the earth's atmosphere and possibly to cool it (though through unacceptable means), we should not rule out that technologies for global cooling, perhaps by injecting the right particulates into the stratosphere, perhaps by subtler means, will become economical during coming decades.

A more benign example, compared with nuclear winter or induced volcanic eruptions, may be the manipulation of cloud cover. Let me again quote Thomas Malone (1968:1135).

> A characteristic of the atmosphere that frustrates the weather forecaster while providing a basis for optimism on the part of the weather modifier is a tendency

> for the processes in the atmosphere to demonstrate certain traits of instability. . . . For example, a small puffy-type cloud may grow to a towering thunderstorm in a matter of hours; a gentle zephyr in tropical latitudes may develop into a "killer" hurricane in a matter of days; and a small low-pressure center may grow to a vigorous extratropical cyclone within a single day. . . . An avenue may be opened up by which great effects may be produced from relatively modest but highly selective human interventions.

If somebody learned in the next 50 years how to affect the extent and global distribution of certain kinds of cloud cover, incoming radiation may become manipulable by nations, international agencies, or even interested private organizations, depending on the nature of the technology, its expense, and perhaps geographical considerations.

It is difficult to mention such a possibility without appearing to recommend it, or to use it as a "technological fix" in the future to divert attention from some need for immediate policy intervention. I am not recommending, I am predicting. Independently of CO_2, we have to consider that weather and climate modification may become feasible in a period of time no longer than the elapsed time since electronics, genetics, antibiotics, and nuclear fission were unimagined. The greenhouse warming may generate an interest among most nations in moderating the changed radiation balance, and if it proves more expensive to facilitate outgoing radiation than to obstruct incoming, there may be powerful motives for considering it. And if the technique for moderating incoming radiation were globally uniform or nearly so, an international agreement would have only to decide how to share the costs, a unidimensional problem compared with sharing the reduction of emissions.

If intervention is more regional than global, or global but not uniform in its distribution, intervention could become exceedingly controversial. Mexico and China are counting on those hurricanes—they are an essential source of rainfall for crops—whereas the Cubans, Filipinos, Japanese, and residents of the Texas coast would suppress them if they knew how.

In closing I must say a word about sea level. I believe the current wisdom is that we may be in for rising sea level that could be on the order of a meter per century for several centuries (Robin, 1986). Anything upwards of a meter, perhaps even half a meter, would primarily be due to the collapse of the West Antarctic Ice Sheet. The full 20-foot rise corresponding to the complete disappearance of that body of ice would put the White House rose garden under water, make Beacon Hill in Boston an island, and isolate the southern third of Florida by making the middle third disappear under water.

A country like the United States should be able to adapt (eventually by doing, perhaps, what the Dutch have been doing for centuries—constructing dikes). No such "easy" solution is available to a country like Bangladesh, which is densely populated in large areas that would be inundated by the full sea level rise, and which could not be protected with dikes. (If dikes were erected along the coastline to protect against seawater flooding, the area would simply be flooded with fresh water that could not flow out to sea.)

If current estimates hold up, the potential devastation of rising sea level will mainly be 100 years away, and the government of Bangladesh should worry much more about population and productivity than climate change. If the more prosperous nations were prepared to help Bangladesh at great expense to themselves, aid now would probably appeal more to Bangladesh than heroic efforts to forestall floods a century hence. (That country already has floods to cope with in this century!)

Estimates of rising sea levels depend not only on thermal warming of the oceans, melting of glaciers, and what happens to the West Antarctic Ice Sheet; they can also depend on what happens to the Antarctic climate. There has been some conjecture that a warming of the South Polar air may lead to greater snowfall on Antarctica. The area of Antarctica is about one-fortieth the area of the oceans; a 1-centimeter rise in ocean level would be offset by a 40-centimeter rise in the water content of the snowfall on Antarctica, or an average snowfall of 4 meters per year. Storing water as ice on Antarctica might be the ideal solution to the water-level problem. Even the people most offended at the thought of deliberately tampering with our climate to offset the greenhouse gases may agree that learning to make it snow on Antarctica is a worthwhile project for the next century.

REFERENCES

Ausubel, J. H., and W. D. Nordhaus. 1983. A review of estimates of future carbon dioxide emissions. Pp. 153–185 in Changing Climate: Report of the Carbon Dioxide Assessment Committee. Washington, D.C.: National Academy Press.

Malone, T. F. 1968. New dimensions of international cooperation in weather analysis and prediction. Bulletin of the American Meteorological Society 49:1134–1140.

National Research Council. 1982. Carbon Dioxide and Climate: A Second Assessment. Carbon Dioxide Review Panel. Washington, D.C.: National Academy Press.

National Research Council. 1983. Changing Climate: Report of the Carbon Dioxide Assessment Committee. Washington, D.C.: National Academy Press.

Nordhaus, W. D., and G. W. Yohe. 1983. Future carbon dioxide emissions from fossil fuels. Pp. 87–152 in Changing Climate: Report of the Carbon Dioxide Assessment Committee. Washington, D.C.: National Academy Press.

Robin, G. deQ. 1986. Changing the sea level: Projecting the rise in sea level caused by warming of the atmosphere. Pp. 323–359 in SCOPE 29: The Greenhouse Effect, Climatic Change, and Ecosystems, B. Bolin, B. R. Döös, J. Jäger, and R. A. Warrick, eds. New York: John Wiley & Sons.

Regional Environmental Forces: A Methodology for Assessment and Prediction

THOMAS E. GRAEDEL

The ongoing impact of human development on the biosphere is a fact of life. Development assumes many different forms, including better housing, improved communications, and advancing technology, ultimately directed at upgrading the quality of human existence. However, human development *is* biospheric development and can ultimately be beneficial only if it is sustainable: that is, if the ways in which development is accomplished do not result in biospheric impacts sufficiently detrimental that they outweigh the benefits pursued.

Among the most significant of the impacts are those resulting from the generation and use of energy. The preponderance of energy generation involves the combustion of fossil fuels, with resulting effects on air, water, vegetation, and so forth. Alternative technologies have detrimental aspects as well, of course. The issues involved are complex, both in the scientific and in the societal sense, and it is this complexity that has led to restricted or simplistic analyses and has hindered development of rational overall plans to optimize energy provisioning.

Throughout most of history, the interactions between human development and the environment have been relatively simple and local. More recently, the interrelatedness and scale of these interactions, as well as our awareness of them, have increased. This increased complexity presents a considerable challenge to those attempting to assess the conditions of the future. It has become clear that environmental impacts and their causes cannot be envisioned in isolation as has traditionally been done, because one impact may have many causes, and one cause may have many effects.

In partial response to these concerns, this chapter presents a methodology for the assessment and prediction of environmental effects of human development. It draws on this author's research experience as well as that of others (especially Clark and Munn, 1986; Darmstadter et al., 1987; Meyer and Turner, 1988). The assessment is limited to the atmosphere and to some of its important interactions with the rest of the planetary system. Regional effects are emphasized, but global effects are considered as well, because the ensemble perspective that is desired must encompass all spatial and temporal scales of interest. A typical regional dimension is defined here as 1,000 kilometers, because most of the world's nation states tend to have spatial dimensions of that order.

The assessments reported here are not intended to be definitive scientific assessments. As Victor Hugo said with considerable discernment, "Science has the first word on everything and the last word on nothing." The research is designed, rather, to suggest a path for the practice of what James Lovelock (1986) has called planetary medicine, or geophysiology. Lovelock's goal is to understand the planetary system well enough to answer such questions as: How stable is it? What will perturb it? How much will it be perturbed? Can the effects of perturbation be reversed? This paper is addressed to providing a framework for answering these questions and thus moving toward developing the biosphere in a rational and knowledgeable manner. To say it another way, the work is intended to assist those desiring to become better planetary physicians.

TOWARD A SYNOPTIC FRAMEWORK

Noteworthy Atmospheric Properties

The goal of a synoptic framework is to establish the causal relationships between atmospheric properties that exert significant influence on an ecosystem, and potential sources of change to these properties (and the processes behind them). The importance of changes in these properties and their consequences for individuals depends on their social, political, and current environmental circumstances. This discussion cannot embrace the full diversity of human interests that could be affected by continued alteration of the atmosphere. However, one of the clearest lessons from the assessment experience of the past decade is that unless some short and clearly defined set of atmospheric "benchmarks" is established, a meaningful analysis is impossible.

Therefore, for the purposes of this discussion, the set of significant atmospheric properties presented in Table 1 is used as a point of departure (Crutzen and Graedel, 1986). The goal is to understand the relationships between these influential properties of the atmosphere, their natural

TABLE 1 Atmospheric Properties and Processes

Property/Process	Description
Ultraviolet energy absorption	The ability of ozone in the stratosphere to absorb ultraviolet solar radiation, thus shielding the earth's surface from its effects.
Radiation balance alteration	The complicated processes through which the atmosphere transmits much of the energy arriving from the sun at visible wavelengths while absorbing much of the energy radiated from the earth at infrared wavelengths. The balance of these fluxes, interacting with the hydrological cycle, exerts considerable influence on the earth's temperature. This process is commonly addressed in discussions of the "greenhouse" problem.
Photochemical smog formation	The oxidizing characteristics of the atmosphere are due to a variety of highly reactive gases. In this paper, the emphasis is on local-scale oxidants that are often implicated in problems of smog, such as asthma, crop damage, and degradation of works of art.
Precipitation acidity	The acid-base balance of the atmosphere in rain, snow, and fog.
Visibility	Visibility is reduced when visible light is absorbed or scattered by gases, moisture, or particles in the atmosphere.
Corrosion potential	The ability of the atmosphere to corrode materials exposed to it, often through the chloridation or sulfurization of marble, masonry, iron, aluminum, copper, and other materials.

fluctuations, and how they might be affected by human activities. Recent advances in our understanding of atmospheric chemistry and its interactions with the biosphere now allow specification of such relationships in terms of fundamental biological, chemical, and physical processes.

Interactions

Current knowledge about the atmospheric properties affected by changes in specific atmospheric chemicals is given qualitative expression in Figure 1. The convention that is used is to indicate only direct effects. For example, changes in ozone concentrations affect ultraviolet energy absorption because the ozone molecules themselves capture the photons, or rays of light. Halocarbons and nitrous oxide, though surely relevant to ultraviolet energy absorption, are not shown to influence this atmospheric property because their effect is indirect. It is important to note from Figure 1 that a significant number of chemicals have multiple impacts. Sources of atmospheric chemicals or policies that affect the introduction of these chemicals into the environment must therefore be assessed in the light of these multiple atmospheric impacts, and the "environmental-problem-of-the-month" approach that has been so common in the past must be avoided.

Figure 2 shows the current state of knowledge about the sources of atmospheric chemicals (see Table 2), again indicating only direct effects. It is again important to note that many atmospheric chemicals have more than one source. The pervasive influence of the biosphere—ocean life, plants, soils, and animals on atmospheric chemistry—is also evident in the figure.

To complete the connection between the sources of atmospheric chemicals and their influence on the important properties of the atmosphere, it is necessary to account for indirect effects—the fact that changes in chemical species A may affect a given atmospheric property through an intermediate influence on chemical species B. Understanding the indirect effects of chemical interactions is one of the central tasks of contemporary atmospheric science. The immense complexity of even the relatively well understood interactions precludes their discussion here. Conceptually, however, the substance of such a discussion can be captured in a matrix constructed along the lines of Figure 3. The chemical compounds of Figures 1 and 2 thus provide the common basis for an analysis of biogeochemical processes that link changes in the contribution of one source of a chemical to all the important atmospheric properties affected.

The three figures can be combined to provide a synoptic framework for atmospheric assessment. One can begin with a property such as "precipitation acidity" and its direct chemical causes (Figure 1), trace those back through their interactions with other atmospheric chemicals (Figure 3), and finally identify the sources of those chemicals that influence precipitation acidity (Figure 2). The synoptic matrix relates the change of each potential source of atmospheric chemical to the affected atmospheric component and can be regarded as being generated by a matrix operation: [source × property] = [source × chemical] [chemical × chemical] [chemical × property], or [Figure 4] = [Figure 2] [Figure 3] [Figure 1]. The initial result of an analysis based on such a matrix, shown in Figure 4, is qualitative, as befits the present state of knowledge. It also includes estimates of the reliability of that knowledge, an important component of such an assessment effort (Ravetz, 1986).

ATMOSPHERIC PROPERTY

ATMOSPHERIC CHEMICAL	Ultraviolet Energy Absorption (by ozone)	Radiation Balance	Photochemical Smog	Precipitation Acidity	Visibility	Corrosion Potential
CO_2		■				
CO						
NO_x			■	■	■	
SO_x		■		■	■	■
H_2S						■
C (soot)		■			■	
O_3	■	■	■			
COS						■
CH_4		■				
C_xH_y			■		■	
N_2O		■				
NH_3/NH_4^+				■		
Organic S						■
Halocarbons		■				
Other Halogens					■	
Trace Elements						

FIGURE 1 Direct effects of atmospheric chemistry on important atmospheric properties. The squares indicate that the listed chemical is expected to have a significant direct impact on the listed property. Definitions of the atmospheric properties are given in Table 1. Data are from sources listed by Clark (1986).

FIGURE 2 Major sources of atmospheric chemicals. The squares indicate that the listed source is believed to exert a significant direct effect on the listed chemical. Definitions of the sources are given in Table 2. Data sources are the same as for Figure 1.

SOURCE	CO_2	CO	NO_x	SO_x	H_2S	C (soot)	O_3	COS	CH_4	C_xH_y	N_2O	NH_3/NH_4^+	Organic S	Halocarbons	Other Halogens	Trace Elements
Oceans and Estuaries	■	■	■		■			■		■	■		■		■	■
Vegetation and Soil	■	■			■			■	■	■	■	■	■			
Wild Animals									■			■				
Wetlands					■			■	■			■	■			
Biomass Combustion	■	■	■			■			■	■	■	■				■
Crop Production	■		■		■				■	■	■	■				
Domestic Animals									■			■				
Petroleum Combustion	■	■	■	■		■					■	■			■	■
Coal Combustion	■	■	■	■		■					■				■	■
Industrial Processes		■	■	■										■	■	■

TABLE 2 Sources of Atmospheric Chemicals

Source	Description
Oceans and estuaries	Includes coastal waters and biological activity of the oceans.
Vegetation and soils	Includes activities of soil microorganisms; does not include wetlands or agricultural systems.
Wild animals	Includes microbes, except for those of soils; does not include domestic and marine animals.
Wetlands	An important subcomponent of vegetation and soils; does not include rice.
Biomass burning	Includes both natural and anthropogenic burning.
Crop production	Includes rice, fertilization, and irrigation, but not forestry.
Domestic animals	Includes grazing systems and the microbial fauna of the guts of domestic animals.
Petroleum combustion	Includes impacts of refining and waste disposal.
Coal combustion	Includes impacts of mining, processing, and waste disposal.
Industrial processes	Includes cement production and the processing of nonfuel minerals and chemicals.

NOTE: See also, Figure 2.

ASSESSMENTS

The simplest atmospheric impact assessments involve only a single cell of the matrix. A typical example is the study of the impacts of a single source, such as a new coal-fired power station, on a single noteworthy atmospheric property, such as precipitation acidity (location a in Figure 4). More complex atmospheric assessments have addressed the question of aggregate impacts across different kinds of sources. A contemporary example is the study of the net impact on the earth's thermal radiation budget caused by chemical perturbations due to fossil fuel combustion, biomass burning, land-use changes, and industrialization (e.g., location d in Figure 4). The assessment then becomes a column total (location e) in the synoptic framework. An example of a "column" assessment is Bolin et al. (1986).

Even more useful for the purposes of policy and management are assessments of the impacts of a single source on several noteworthy atmospheric properties. The simple study noted above would fall into this category if it assessed the impacts of coal combustion not only on acidification but also on photochemical oxidant production, materials corrosion, visibility, the radiation balance, and stratospheric ozone (i.e., locations b in Figure 4). The impact assessment then becomes a row total of the

Impact of \ Impact on	CO_2	CO	NO_x	SO_x	H_2S	C (soot)	O_3	COS	CH_4	C_xH_y	N_2O	NH_3/NH_4^+	Organic S	Halocarbons	Other Halogens	Trace Elements
CO_2																
CO																
NO_x																
SO_x																
H_2S																
C (soot)																
O_3																
COS																
CH_4																
C_xH_y																
N_2O																
NH_3/NH_4^+																
Organic S																
Halocarbons							*									
Other Halogens																
Trace Elements																

FIGURE 3 Atmospheric chemical interaction matrix for the compounds listed in Figures 1 and 2. As an example of the matrix element entries needed to fill this matrix, the element indicated by the asterisk represents an assessment of how changes in halocarbons affect ozone concentrations.

synoptic assessment matrix (location c). Examples of "row" assessments are National Research Council (1979, 1981).

Figure 4 shows that the sources of most general concern, as indicated by their impact ratings, are almost wholly anthropogenic: fossil fuel combustion, biomass combustion, and industrial processes. Emissions from crop production, especially methane from rice paddies, and from estuaries near heavily populated areas may have future impacts. Some sources (the animal kingdom and vegetation) have sufficiently small effects on atmospheric processes that they need not cause much concern, even if their emissions should increase.

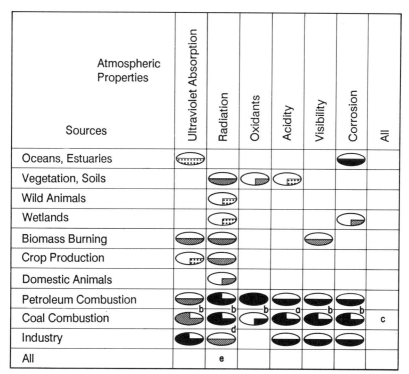

FIGURE 4 A synoptic assessment of impacts on the atmosphere. In this figure, adapted from Clark (1986), the atmospheric properties defined in Table 1 are listed as the column headings of the matrix. The sources of disturbances to these properties as defined in Table 2 are listed as row headings. Cell entries assess the relative impact of each source on each component and the relative scientific certainty of the assessment. "Column totals" would, in principle, represent the net effect of all sources on each noteworthy atmospheric property. "Row totals" would indicate the net effect of each source on all noteworthy atmospheric properties. These totals are envisioned as judgmental, qualitative assessments rather than as literal, quantitative summations. The cells labeled a through e are discussed in the text.

Figure 4 also suggests areas in which useful interdisciplinary studies could be performed. These are areas that have considerable potential importance but whose assessment reliability is low to moderate. The two best candidates for such studies are biomass burning (thought to play an important role in tropical photochemistry, the atmospheric radiation balance, and visibility) and vegetative emissions (potentially linked to photochemical oxidant formation, the radiation balance, and precipitation acidity).

The diversity of primary anthropogenic sources of causative chemical species shown in Figure 4 is surprisingly small: fossil and biofuel combustion and industrial processes being dominant. There is enough evidence to state that a significant decrease in the rate of fossil fuel combustion would tend to stabilize the atmospheric radiation balance, improve visibility, hinder smog formation, and minimize acidic precipitation and its effects. Stratospheric modification is best constrained by devising alternatives to the use of chlorofluorocarbons (CFCs) as aerosol propellants and refrigerants. The involvement of transition metal chemistry in droplets and aerosol particles is not yet understood, but better control of metal emission from combustion and smelting operations may be desirable to alleviate acid deposition problems. The problem of corrosion would have to be handled differently, because natural sources are responsible for many of the corrosive agents. Thus, it might be most effective to reduce corrosion by improved selection and treatment of materials rather than by control of emission sources.

EXTENDING THE FRAMEWORK IN SPACE AND TIME

Concepts and Goals

Thus far, the synoptic framework developed in the previous section has only considered interactions occurring at a given instant in time. However, this is insufficient because many of the impacts of human development on the environment are cumulative. This arises because many sources of atmospheric disturbance operate over space and time scales such that the disturbing species can accumulate faster than it is removed. Although the interplay is complex, to a first approximation, the longer a species remains in the atmosphere, the more likely it is to accumulate and the greater its spatial impact can be. For example, heavy hydrocarbons and coarse particles are short lived, dropping out of the atmosphere in a matter of hours. Hence, they have little opportunity to travel more than a few hundred kilometers from their sources. On the other extreme, the residence time of carbon dioxide (CO_2) in the atmosphere is so long that production of this gas anywhere in the world contributes to its effects everywhere in the world. This is why the greenhouse effect has its long-term, global-scale character. The species with moderate atmospheric lifetimes include a group of chemicals associated with the acidification of precipitation. Because these substances last a few days, they have the opportunity to affect regions as far away as a 1,000 kilometers or more from the source of the emission.

These concerns, together with the evidence that the concentrations of many chemical compounds in the biosphere are increasing, indicate that additional extensions to the conceptual framework are needed to provide it with spatial and temporal dimensions. The effort described earlier focused on present-day impacts across various local, regional, and global conditions. One can envision preparing separate versions of the framework shown in Figure 5 for global, regional, and local interactions, and for different epochs in time. The assessment should eventually consist of a single, global-scale analysis, plus several analyses for specific, large-scale regions (e.g., Europe) selected to reflect interesting interactions between development and environment.

For each spatially defined regime, a sequence of figures would be needed to show how the relations between sources and noteworthy environmental properties change through time. As suggested in Figure 5, this sequence might consist of separate versions of Figure 4 created to reflect "slices in time" through the evolving conditions at 25- or 50-year intervals. In the version being explored, this sequence will extend several hundred years into the past and a century into the future. The result will help to put the changing character of interactions between human activities and the environment into a truly synoptic historical and geographical perspective. This historical dimension requires assessment of the kind of human activity being carried out at specific times in the past, and how much and what kind of activity might occur in the future.

Air Quality Assessment and Scenario Design

The degree of development of a country or geographic region has a major influence on its air quality, and the intensity of technology and of land use in a geographical region define many of that region's impacts on the atmosphere and therefore must be taken into account. It is often possible to locate information about the most frequently monitored air-quality parameters in the major cities of the world. Information is sometimes available also for less widely reported atmospheric constituents or for other locations, although such information is generally unearthed only through informal contacts and negotiated exchanges. Where no measurements have

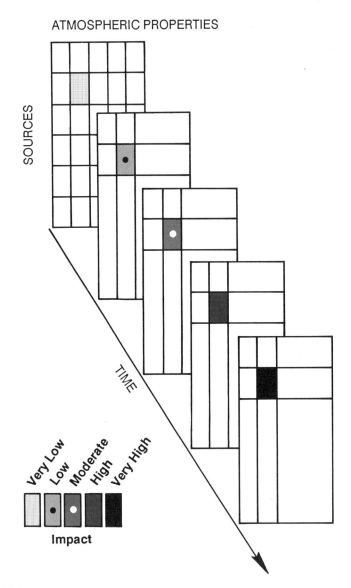

FIGURE 5 A history of disturbances to the atmosphere, as might be expressed through a time series of source-impact matrices such as that shown in Figure 4. This display suggests how one of the matrix elements would be evaluated at each of the time slices, perhaps becoming more significant with time as indicated here by the shading of the symbol. The full assessment would include such an evaluation for each of the individual matrix elements at each time slice.

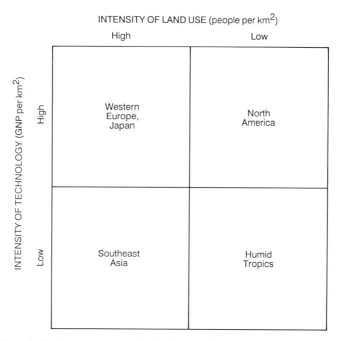

FIGURE 6 A development matrix for the interaction between intensity of land use and intensity of technology. Regions typical of each matrix element are listed. SOURCE: After Crosson (1986).

been made, information on emissions sources such as power plants, industrial processes, and transportation facilities can often suffice to give initial estimates of air quality.

As noted, human response to atmospheric impacts depends, in large part, on the current developmental and environmental situations that prevail in a given region. For this reason, it is useful to categorize regions on the basis of land use and energy intensity. Figure 6 shows simple representation of divisions into which the earth may be divided by these parameters (Crosson, 1986). Darmstadter et al. (1987) have shown that it is possible to project future air quality for regions typical of each cell in the Figure 6 matrix:

- Western Europe (high intensity of both land use and technology)
- Northeastern North America (high technology, moderate land use)
- India's Gangetic Plain (high land use, low technology)
- Amazonia (low land use, low technology)

This straightforward classification scheme embraces some obvious relationships: that a high intensity of land use implies impacts on soil and groundwater chemistry. Perhaps less obvious are the spatial scales of the impacts, which tend to be large for high-technology regions and small for low-technology regions.

It is beyond the scope of this chapter to discuss examples of assessment in each of the regions of Figure 6. Europe is an appropriate region to show the process, however, because the high density of technology and population in the region has resulted in readily identifiable impacts. It is an appropriate region also because it may serve as a predictor of trends in regions at lower levels of environmental loading, if development in those regions proceeds along lines similar to those of Europe.

Two development scenarios were used in the Darmstadter study; they can be summarized briefly as follows. The first (referred to as scenario C in reference to its constant emission coefficients) is a conventional growth scenario for the world. It predicts global growth rates of 1.5, 3.2, and 2.1 percent per year for population, gross national product, and total energy use, respectively, for the period 1980–2030. For the period 2030–2080 the comparable growth rates predicted are 0.3, 2.2, and 1.8 percent per year. For each period, appropriate growth rates are assigned also to particular geographical regions of interest. The emission factors (i.e., the mass of emittants per unit mass of feedstock in a given source process) are assumed to remain constant at 1980 values. Scenario C is regarded as a conservative but not particularly restrictive projection for world development.

The second growth scenario (termed scenario D in reference to its decreasing emission coefficients) is identical to the first in its global and regional growth rates. It differs from scenario C by assuming emission factors for the sources will decrease over the century 1980–2080 at a rate of 1 percent per year. For this scenario, therefore, increases in the total use of feedstocks are offset somewhat by improvements in technology. Scenario D is more conservative than scenario C in its potential atmospheric impacts.

Given these scenarios, it is possible to use a variety of computer models and evaluation techniques to derive predicted concentrations of selected atmospheric parameters. In the study by Darmstadter et al. (1987), four noteworthy atmospheric properties, photochemical smog, acid precipitation, corrosion of metals, and stratospheric ozone were selected for analysis for the epochs 1890, 1920, 1950, 1980, 2030, and 2080. The results for Europe are reported in Table 3.

Assessment for Europe

The "Europe" study region used by Darmstadter et al. (1987) comprises all of Western Europe. Heavy industry occurs throughout the region, with several localities of intense industrialization. The density of motor vehicles is high, particularly in and near the major urban centers. Two urban areas

TABLE 3 Environmental Quality Assessments for Europe

	Epoch							
	1890	1920	1950	1980	2030C	2080C	2030D	2080D
O_3, ppb	10.0	20.0	35.0	55.0	180.0	280.0	130.0	130.0
Smog severity	L	L	L/M	M	H	H	H	H
Precipitation pH	5.4	5.4	5.3	4.3	3.5	2.8	3.8	3.3
Precipitation acidity	L	L	L	M	H	H	H	H
SO_2, ppb	7.0	11.0	12.0	14.0	40.0	160.0	25.0	60.0
Cl^-, µeq/l	30.0	37.0	39.0	42.0	62.0	120.0	54.0	72.0
Corrosion severity	L/M	M	M	M	M/H	H	M	M/H

KEY: C—Constant-emissions scenario; D—declining-emissions scenario; L—low impact; M—moderate impact; and H—high impact.

SOURCE: Darmstadter et al. (1987).

serve to characterize the region. One is Brussels, a city of 2.2 million people, generally upwind of the principal industrial centers of the region; the other is Stockholm, a city of 1.6 million people, which is generally downwind of emission sources.

Although variations in topography exert influences on local weather conditions, a principal feature of the weather pattern is its consistency for all of Europe. The pattern characterized by unsettled weather throughout the year as a consequence of the migration of weather systems from west to east. Sometimes, however, a region of low or high pressure can be stationary over a wide area, resulting in uniform weather there for several days or even weeks. Such conditions promote the buildup of atmospheric contaminants.

Air quality measurements in Europe have been taken for about two decades, but extensive data are available only from about 1975. The 1980 epoch can thus be readily defined, and it can be related in some detail to current emissions inventories, especially those for sulfur dioxide (SO_2). Emissions data for nitrogen dioxide (NO_2) are less complete, although the need for such data is minimized somewhat by the extensive ozone (O_3) measurements that can be drawn upon. Selected measurements of other airborne trace species are available as well. In the absence of historic measurement data, estimates of the atmospheric properties can generally be made by comparing human development activities at earlier times with those of the present day and inferring emission fluxes from those comparisons.

To project atmospheric properties over future epochs, plausible models for development must be formulated to estimate the impacts of biospheric development. These models are initialized and validated by the historical information that is available. The development scenarios selected for Europe call for increases of 0.5, 2.4, and 1.8 percent per year, respectively, in population, gross national product, and total energy use for the period 1980–2030. For the period 2030–2080, the corresponding rates are 0.3, 2.0, and 1.5 percent per year. In nearly every case, these numbers are significantly lower than the estimated global average values; for example, Europe is expected to undergo development during the next century at rates lower than those for most of the rest of the world.

Regional Impacts

Photochemical Smog Although photochemical smog consists of many trace species, ground-level ozone is its best single indicator. Bojkov (1986) and Volz and Kley (1988) have reevaluated and summarized the surprisingly extensive ozone data from a number of stations in Europe during the interval 1850–1900. The ozone concentrations during that period are about one-fourth of the mean of the daily maximum values of precise surface ozone measurements taken in the same geographical regions during the past 10–15 years, indicating that significant increases have occurred. For 1980, measured concentrations establish the average European ozone value. The concentrations during the epochs between 1900 and 1980 are established primarily on the basis of the trends in emissions of the oxides of nitrogen (NO_x). Table 3 indicates that for scenario C, the predicted ozone concentrations are quite high, the values being comparable to those of the Los Angeles basin on a smoggy day. A more moderate increase in smog concentrations is predicted for scenario D.

Acid Precipitation Data on precipitation chemistry take many forms, but the acidity, or pH, is perhaps the most meaningful. A few historical measurements of precipitation chemistry in Europe are useful. For the mid-1950s the concentrations of sulfate and nitrate in precipitation were determined in Stockholm (Environment '82 Committee, 1982). Some historical data on acidity of precipitation also exist for several locations in Europe, although data on sulfate ($SO_4^=$) and nitrate concentrations are thought to be more reliable. Contour maps of precipitation acidity in Europe for the past several decades are presented by Likens et al. (1979).

Precipitation chemistry in earlier epochs can be estimated by relating its characteristics to that of known levels of emission of precursor species.

The data described above establish the values of acidity or related parameters at several epochs. Emission fluxes and relational evaluation techniques can then be used to estimate acidities for the development scenarios. As shown in Table 3, acidity of precipitation was generally of little concern before 1950, when it began to increase rapidly. The projections for the future, by either scenario C or scenario D, indicate that high acidity of precipitation is envisioned for the twenty-first century.

Corrosion of Metals Because the corrosion of metals exposed to the atmosphere nearly always involves chlorine and sulfur, it is appropriate to study common atmospheric forms of those elements. Given 1980 data, one can derive historical $SO_4^=$ and Cl^- concentrations and proceed to make corrosion assessments. Precipitation data for sulfate in the early 1950s allow a good assessment for that epoch as well. An assessment of corrosion potential in Europe is then derived by relating the precipitation $SO_4^=$ and Cl^- concentrations to the European fluxes of SO_2 to the atmosphere. The results, shown in Table 3, indicate that the corrosion impact has been low to moderate since the midpoint of the current century. The projection for development scenario C (constant emissions) is that atmospheric corrosion during the twenty-first century will be very severe. Under scenario D (declining emissions), the corrosion impact is slightly less severe but still worthy of concern.

Global Impacts

Ultraviolet Absorption Europe is a major producer of CFCs. As such, it bears a substantial responsibility for any stratospheric ozone depletion that occurs, and hence for decreases in the ultraviolet absorption capacity of the atmosphere. Under scenario C, the impact attributed to Europe is severe. Under scenario D, improvements in technology and reductions in manufacturing levels modify that impact somewhat, but it is still moderate to moderately severe throughout the period included in the present study. The study did not include the emission controls that may result from the 1987 multinational Montreal Protocol on Substances That Deplete the Ozone Layer, which, if implemented, would reduce the impact of the assessments.

Atmospheric Heat Retention The heat retention propensity of the atmosphere, that is, the potential for the greenhouse effect, is directly related to the atmospheric concentration of carbon dioxide and other gases capable of absorbing the outgoing radiation from the earth's surface.

Changing carbon dioxide scenarios were not studied by Darmstadter et al. (1987) but have been widely investigated elsewhere.

A common calculation for the study of atmospheric heat retention is one in which the concentration of atmospheric carbon dioxide is set to twice its current value. (It is predicted that such concentrations will be reached at some point in the twenty-first century.) The results of calculations by several groups of researchers have been presented and discussed by Luther (1985) and Grotch (1988), and some of their comments are abstracted here.

For doubling of the CO_2 concentration, the change in global mean surface air temperature ranges in different models from 1.5 to 4.5°C. The increase is not uniform but is generally 1–3°C near the equator and 4–16°C at high northern latitudes during winter months. This increase in temperature will be sufficient to reduce snow cover, melt some sea ice, and strongly influence the global water cycle. The models predict warming in the troposphere and cooling in the stratosphere, with somewhat unpredictable impacts on the global circulation patterns. Cloud cover and precipitation will change as well, though the ability of researchers to model these changes is somewhat uncertain. All of the models predict an increase in global mean precipitation (note that a warmer atmosphere can hold and thus cycle more water than a cooler one). Precipitation decreases will probably occur in some areas of the globe, however.

It is difficult to predict the environmental impacts of increased atmospheric heat retention in specific regions, especially small ones, since model simulations do not agree well on length scales less than a few thousand kilometers. Assigning responsibility for the global impacts is easier, because the causative agents are identified: carbon dioxide and, to a lesser extent, methane, nitrous oxide, ozone, and a variety of chlorofluorocarbons. Fossil fuel combustion is the dominant source of CO_2 and nitrous oxide as well as the principal source of the precursors of tropospheric ozone and a minor source of methane (Wuebbles and Edmonds, 1988). Allocation of the global effects of atmospheric heat retention to specific regions can thus be made on the basis of each region's use of fossil fuels and chorofluorocarbons.

Summary

As shown in Table 3, most air quality indicators that were satisfactory near the end of the nineteenth century are noticeably degraded today and can be expected to deteriorate markedly from present levels over the century to come. Given the recognition of a problem, the next information requirement is to identify the factor or factors causing that problem. Identification of this type is the purpose of Figures 7 and 8. These figures follow the intent of the "time slice" diagrams of Figure 5, but use clustered slice rectangles to condense six Figure 5 style diagrams into a single display.

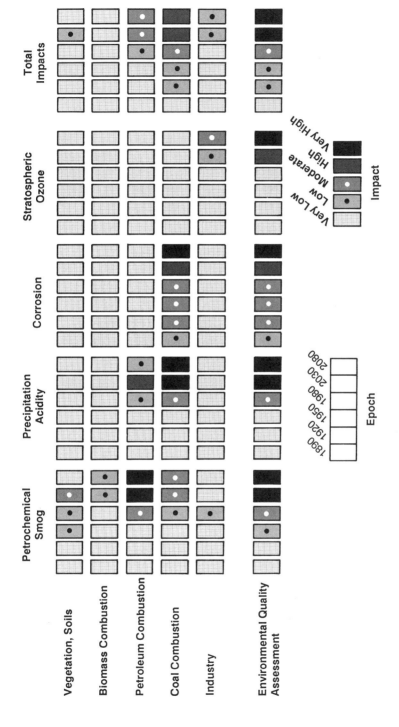

FIGURE 7 Source-impact assessment for Western Europe for scenario C. The details of the techniques used in the construction of this diagram and that of Figure 8 are described in the text and by Darmstadter et al. (1987).

On these figures, a qualitative source impact assessment is shown for each epoch for the several environmental impacts produced by each source type.

It is instructive first to examine Figure 7 in detail, which treats impacts for the past century in the context of scenario C. The bottom row in Figure 7 is a coded version of the environmental quality assessments given in Table 3 (e.g., it contains the column summations of the impacts). The display demonstrates that few measurable impacts occurred before 1980 and that only the combustion of coal and petroleum cause major environmental impacts at present. For the future, the same sources are expected to be those primarily responsible for the deterioration in regional air quality, with industrial emission of CFCs contributing to stratospheric ozone reduction. Biomass combustion and emission from vegetation and soils will play small roles, especially in photochemical smog production.

The source impact assessment for Europe under scenario D is shown in Figure 8. It resembles that of Scenario C except for modest decreases in the severity of the impacts, reflecting the assumption of declining emission coefficients. Even with this assumption, however, the impact on precipitation acidity remains very high in both 2030 and 2080; the impact on smog and corrosion improves slightly but remains high. The ensemble impact is rated high in 2030 and very high in 2080. Thus, even the unremarkable development postulated in scenario D and the assumed significant decline in emission coefficients (more than 60 percent) leave the region with major atmospheric perturbations a century from now. As in scenario C, coal and petroleum combustion and CFCs are the sources primarily responsible for this deteriorating condition.

Since Figures 7 and 8 (and similar figures can be produced for other geographical regions) serve as summaries of the assessment efforts, it is appropriate to call attention to certain aspects of the diagrams. One point is that smog, acidity, and corrosion are local or regional in scale and are largely a consequence of emissions from within the region. Stratospheric ozone depletion, however, has been assessed on a global basis and the indication in the diagram refers to a regional contribution to that global impact. Strict parallelism is thus sacrificed to the desire to communicate a more comprehensive view of the effects of emissions. A second point with respect to Figures 7 and 8 is that the four atmospheric perturbations are assigned equal value in determining the total impact from each of the sources. Whether different weights should be assigned to the perturbations is a policy question, not an analytic one; Darmstadter et al. (1987) and the extensions to it considered here adopted the simplest of the possible choices.

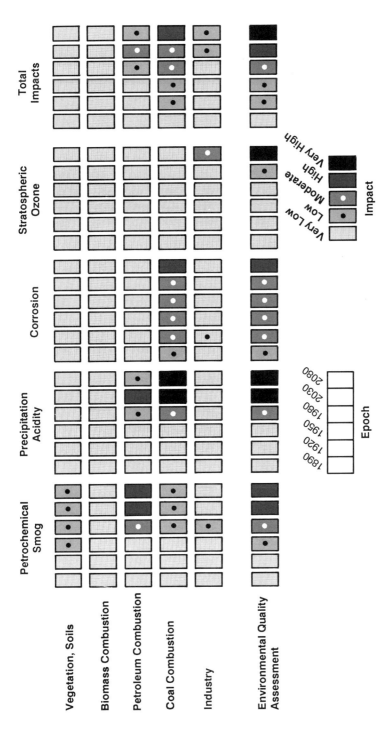

FIGURE 8 Source impact assessment for Western Europe for scenario D.

EXTENDING THE FRAMEWORK TO DIFFERENT REGIMES

It is important to recall at this point that the previously presented assessments have been limited to the atmospheric effects of development. In fact, other regimes in the biospheric system are important as well. For example, concern about the cumulative effects of acidic precipitation is based not on the accumulation of certain chemicals in the atmosphere, but rather on the accumulation in other media, such as soil and water, and on the increasing fluxes of precursors of acidic precipitation as a consequence of the growing use of fossil fuels. Expanding the synoptic framework to contend with these additional environmental and developmental dimensions is a major part of research on sustainable development. Beginnings have been made for water systems (Douglas, 1976) and soil systems (Harnoz, 1988).

A relatively easy addition to the framework depicted in Figures 1–4 would be one or more noteworthy environmental properties that reflect the role of atmospheric chemicals as direct fertilizers or toxins for plants. Such a modification would allow the integrated treatment of such phenomena as the stimulation of plant growth by carbon dioxide and its inhibition by sulfur oxides—both products of fossil fuel combustion. Somewhat more ambitiously, the approach could be expanded beyond its present chemical focus to include the appropriate physical and biological processes and the sources of disturbance to them. Dickinson's (1986) sketch of a comprehensive framework for understanding the impact of human activities on climate shows the potential of such an integrated approach. Ultimately, the need is for a qualitative framework that puts in perspective the impacts of human activities and natural fluctuations, not just on the atmospheric environment, but also on soils, water, and the biosphere as a whole.

Although a comprehensive treatment along the lines of Figures 8 and 9 has yet to be made for other regimes, it is possible to envision one way in which such an assessment might be accomplished. The first step would be to select regimes of interest other than the atmosphere, such as oceans or soils. The second step would be to use the same sequence used for the atmosphere in Figures 1–4 to determine the matrix elements on a source-impact diagram similar to that of Figure 4. A possible example for soils is shown in Figure 9. The matrix elements are then summed to give a total regime impact for each source; these appear in the right column of Figure 9. The same sequence of operations is repeated for each regime of interest, not only for the present and perhaps for the past, but for future times as well.

In the final ensemble display, the total impact columns from each regime display are extracted and combined, producing a display of the

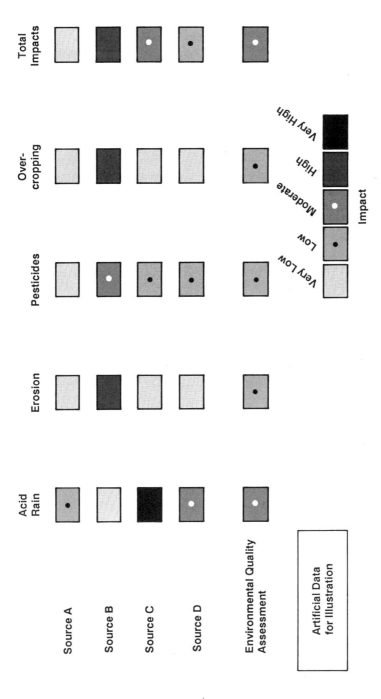

FIGURE 9 A hypothetical synoptic assessment of soil impact analogous to the atmospheric impact assessment shown in Figure 4.

108

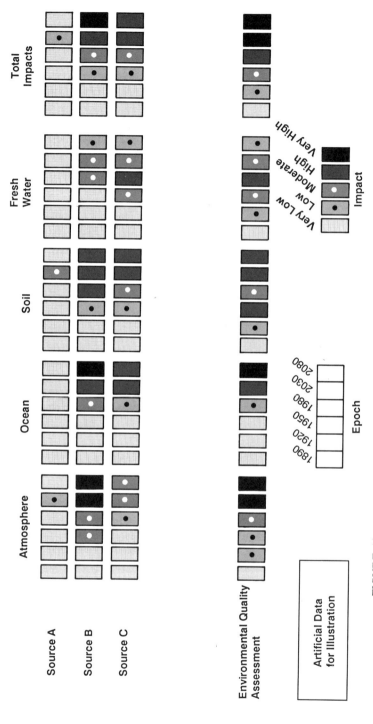

FIGURE 10 A hypothetical source-impact assessment for different parts of the biosphere over the epochs shown in Figure 7.

type shown in Figure 10. A sequence of such diagrams would be used to illustrate progress or retrogression over time.

DISCUSSION AND CONCLUSIONS

Human activities are now major forces affecting the evolution of our planet. These forces have impacts on the environment on many different time and space scales. On a local spatial scale, concern about urban carbon monoxide concentrations is important. On a regional scale, the health of a major river may be of concern. On a global scale, the stability of the climate cycle is a central question. Current assessments of all of these impacts are important, as is prediction of future impacts.

Few environmental impacts have a single cause, and few causes have a single effect. If problems are examined one at a time, a remedy for one problem may exacerbate another. The planetary system is complex but not incapable of being studied in ensemble form if useful techniques for doing so are developed and applied to the problem. It is, of course, much easier to provide a framework of unfilled matrix elements for environmental assessment than it is to fill the elements with assessments or projections made by highly trained and highly motivated scientists, but only by accomplishing such assessments and projections can environmental problems be viewed from a rational perspective.

The methodology presented in this chapter for assessment and prediction of the consequences of environmental forces is primarily applicable to atmospheric impacts at the regional scale, although the approach can be extended to other environmental regimes and time and space scales. It is obvious that global demand for supplies of energy is increasing, and it is equally obvious that the environmental performance of the technologies used to provide that energy will be crucial to the sustainability of the planet itself. Informed decisions regarding impacts and their causative technologies are crucial; they can be well informed only if the information on which they are based is comprehensive, expertly evaluated, clearly presented, and fully utilized.

ACKNOWLEDGMENT

I thank W. C. Clark for insightful comment and continuing encouragement.

REFERENCES

Bojkov, R. D. 1986. Surface ozone during the second half of the nineteenth century. Journal of Climate and Applied Meteorology 25:343–352.

Bolin, B., B. R. Döös, J. Jäger, and R.A. Warrick. 1986. The Greenhouse Effect, Climatic Change, and Ecosystems, SCOPE 29. New York: John Wiley and Sons.

Clark, W. C. 1986. Sustainable development of the biosphere: Themes for a research program. Pp. 5–48 in Sustainable Development of the Biosphere, W. C. Clark and R. E. Munn, eds. Cambridge, England: Cambridge University Press.

Clark, W. C., and R. E. Munn, eds. 1986. Sustainable Development of the Biosphere. Cambridge, England: Cambridge University Press.

Crosson, P. 1986. Agricultural development—Looking to the future. Pp. 104–136 in Sustainable Development of the Biosphere, W. C. Clark and R. E. Munn, eds. Cambridge, England: Cambridge University Press.

Crutzen, P. J., and T. E. Graedel. 1986. The role of atmospheric chemistry in environment-development interactions. Pp. 213–250 in Sustainable Development of the Biosphere, W. C. Clark and R. E. Munn, eds. Cambridge, England: Cambridge University Press.

Darmstadter, J., L. W. Ayres, W. C. Clark, P. Crosson, P. J. Crutzen, T. E. Graedel, R. McGill, J. F. Richards, and J. A. Tarr. 1987. Impacts of World Development on Selected Characteristics of the Atmosphere: An Integrative Approach. ORNL/Sub/86-22033/1. Oak Ridge, Tenn.: Oak Ridge National Laboratory.

Dickinson, R. E. 1986. Impact of human activities on climate—A framework. Pp. 252–289 in Sustainable Development of the Biosphere, W. C. Clark and R. E. Munn, eds. Cambridge, England: Cambridge University Press.

Douglas, I. 1976. Urban hydrology. Geographical Journal 142:65–72.

Environment '82 Committee. 1982. Acidification today and tomorrow. Translated by S. Harper. Swedish Ministry of Agriculture, Stockholm.

Grotch, S. L. 1988. Regional Intercomparisons of General Circulation Model Predictions and Historical Climate Data. DOE/NBB-0084. Washington, D.C.: Office of Energy Research, U.S. Department of Energy.

Harnoz, Z. 1988. The Role of Soils in Sustainable Development of the Biosphere. Laxenburg, Austria: International Institute for Applied Systems Analysis.

Likens, G. E., T. J. Butler, J. N. Galloway, and R. F. Wright. 1979. Acid rain. Scientific American 241(4):43–51.

Lovelock, J. E. 1986. Geophysiology: A new look at earth science. Bulletin of the American Meteorological Society 67:392–397.

Luther, F. M. 1985. Projecting the climatic effects of increasing carbon dioxide: Volume summary. In Projecting the Climatic Effects of Increasing Carbon Dioxide. DOE/ER-0237. Washington, D.C.: Office of Energy Research, U.S. Department of Energy.

Meyer, W., and B. L. Turner, eds. 1989. The Earth as Transformed by Human Action. Cambridge, England: Cambridge University Press.

National Research Council. 1979. Alternative Energy Demand Futures. Committee on Nuclear and Alternative Energy Systems. Washington, D.C.: National Academy of Sciences.

National Research Council. 1981. Atmosphere-Biosphere Interactions: Towards a Better Understanding of Ecological Consequences of Fossil Fuel Combustion. Washington, D.C.: National Academy Press.

Ravetz, J. R. 1986. Usable knowledge, usable ignorance: Incomplete science with policy implications. Pp. 415–432 in Sustainable Development of the Biosphere, W. C. Clark and R. E. Munn, eds. Cambridge, England: Cambridge University Press.

Volz, A., and D. Kley. 1988. Evaluation of the Montsouris series of ozone measurements made in the nineteenth century. Nature 332:240–242.

Wuebbles, D. J., and J. Edmonds. 1988. A Primer on Greenhouse Gases. DOE/NBB-0083. Washington, D.C.: Office of Energy Research, U.S. Department of Energy.

Energy: Production, Consumption, and Consequences. 1990.
Pp. 111–142. Washington, D.C.:
National Academy Press.

The Automobile and the Atmosphere

JOHN W. SHILLER

Many of the necessities and comforts of our daily life are made possible by motor vehicles, which contribute not only to the creation and delivery of the essentials of life but also to the privilege of expanded personal mobility. In urban areas both the favorable and the unfavorable environmental aspects of the energy sources selected to operate motor vehicles are important determinants of our quality of life. The urban environmental factors associated with motor vehicle use are the subject of this chapter.

As with any business, the motor vehicle industry must be responsive to both customer and societal needs. For example, the balance of customer needs and expectations, resource availability, and governmental and societal requirements of today's transportation system represents the dynamic culmination of many competing innovations and options that have been selected and tested with the passage of time. From many points of view, our current motor vehicle transportation fleet represents a good balance of values, including cost efficiency, safety, environmental needs, and the direct requirements and desires of vehicle owners. Any energy strategy designed to turn away from petroleum, which supplies more than 97 percent of the energy needed to operate our transportation system, will require a new balance. If this is to occur, the challenge will be to find a balance that continues to provide value to the consumer as well as to society as a whole.

ALTERNATIVE MOTOR VEHICLE ENERGY STRATEGIES

Consideration of a wide variety of energy strategies started at the very beginning of the development of motor vehicles. The early days of the motor vehicle era were characterized by no firm ground rules or guidelines on which to base vehicle design and associated energy strategy. The fathers of the automobile nudged knowledge forward in tentative bits and pieces. Many of their early ideas are recorded in the automotive magazine *The Horseless Age*, a monthly journal that was devoted to motor vehicle interests during their early development (ca. 1895–1918). It is interesting to note the rich variety of energy strategies reported, including motor vehicles built to operate on gunpowder, calcium carbide-acetylene, compressed air, ether, compressed springs, carbonic acid-sestalit, electric battery, alcohols, coal gas, crude petroleum, gasoline, and other energy sources (Bolt, 1980; Hagen, 1977; Ingersoll, 1895, 1897a–e; Staner, 1905). During the development of these ideas and with advancements in both engine technology and fuels, the advantages of liquid fuels slowly became apparent in such important practical considerations as driving range, convenience, safety, and cost.

The Energy Storage Qualities of Liquid Fuels

As an illustration of one of the advantages of liquid fuels, Figure 1 shows that the specific energy of liquid fuels far exceeds that of alternative gas, solid, or electric fuel systems. Specific energy is defined as the energy available to drive a motor vehicle per unit mass of fuel, fuel container, and fuel transfer equipment. The figure is approximate because the relative ranking of energy strategies, in terms of specific energy, depends not only on the state of technological development but also on other considerations such as safety and cost. A low value for specific energy means a more limited driving range before the need to refuel, which can be offset to some extent if one is willing to use a more massive on-board energy storage system (limited by vehicle design, function, and size) or if less conservative safety factors are used (such as operating closer to the rated bursting pressure of compressed gas tanks).

In Figure 1, gasoline is used as the basis for comparison by normalizing its specific energy to 1.0. For current gasoline vehicles, the specific energy is the amount of energy available to drive the vehicle per unit mass of the fuel, fuel container, and transfer equipment, such as a common fuel tank and pump. For electric vehicles, it is the amount of energy per unit mass of the battery and electrical cables. Thus, the storage mass effectively is constant across all scenarios in the figure. Because specific energy relates to available driving energy, typical engine and power train conversion efficiencies (useful mechanical work output divided by energy input) have

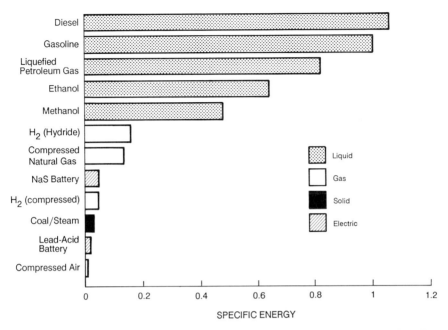

FIGURE 1 Specific energy for various alternative energy systems. In this comparison the relative value for gasoline is set to 1.0.

been included. The available energy of the stored fuel is based on published heat of combustion data (not including the heat of vaporization of water) or its equivalent (Chemical Rubber Company, 1983). The energy required to manufacture each fuel or energy source is not included. This figure shows that the amount of useful mechanical energy available to operate current-technology motor vehicles, at a combined fixed mass of fuel and its on-board storage system (relative to gasoline), varies widely for these different energy strategies and is greatest for liquid fuels.

The alternative energy form closest in performance to petroleum-based fuels is the alcohol fuels: ethanol and methanol. If fluctuations in crude oil prices or availability and changes in U.S. willingness to depend on imported oil require a move away from petroleum, alcohol fuels are an alternative worth considering. Another reason for interest in alcohol fuels is their possible (but not proven) potential for emission reductions that may help attain Clean Air Act air quality standards.

The promotion of alcohol fuels is not new; and with one exception (Brazil), alcohol has failed to gain significant market share. In 1925, Henry Ford predicted that the fuel of the future would come from agricultural products. Consistent with that prediction, in 1927 the first Ford Model A vehicles that were manufactured came equipped with a manually adjustable carburetor specifically intended to give the vehicle owner a choice of using either gasoline or ethanol. Today, about 60 years later, the adjustment to accommodate alcohol fuels, gasoline, or any mixture of the two is accomplished automatically with Ford's experimental flexible-fuel vehicle (Nichols, 1986). Fuel flexibility reduces the risk of investing in a vehicle capable of running on alcohol because a switch to gasoline can be made easily at any time. Beginning in 1926 and continuing throughout the 1930s, William Jay Hale, a chemist at Dow Chemical Company, lauded the benefits of alcohol fuel and started the "power alcohol" promotional movement in the United States. Other countries were also involved in alcohol motor fuels. Brazil passed a law in 1931 requiring gasoline importers to mix their fuel with domestic ethanol made from sugar cane and in 1979 launched an extensive program for vehicles fueled by neat (100 percent) ethanol in addition to its continuing program on alcohol-gasoline blends (Bernton, 1979; Szwarc and Branco, 1987). Other countries, including Chile, China, Czechoslovakia, Germany, Hungary, Sweden, Austria, France, Italy, and Poland, also had some form of alcohol-gasoline blending program after 1930 (Bernton, 1979).

History suggests that the optimum energy strategy for motor vehicle operation will continue to evolve as long as customer and societal values, resource availability or cost, and technological knowledge are subject to change. To date, the relatively low cost of gasoline and diesel fuel and their desirable inherent advantages have preempted the use of alternative energy systems.

DOMINANCE OF GASOLINE AND DIESEL FUEL

Virtually all the oil produced in the 1800s was used to meet the demand for kerosene (a good substitute for whale lamp oil), lubricants, and wax. Gasoline was considered a nuisance and was run off into rivers or flared because no significant market for the product existed until the development of the internal combustion engine in the early 1900s.

With the growth of the nation, total U.S. energy consumption reached nearly 77 quads (quadrillion Btu) in 1987 (U.S. Energy Information Administration, 1988). Figure 2 shows that from 1955 to 1986, petroleum was the largest and the fastest growing segment of total energy consumption. In 1986, 43 percent of total U.S. energy consumed was supplied by petroleum. Highway vehicles consumed nearly 50 percent of the petroleum or 21 percent of the total energy used in the United States, as shown in Figure 3. U.S. domestic production of petroleum has not kept pace with consumption, as shown in Figure 4.

There are many reasons for the dominance of oil in the U.S. transportation sector. For most of our history, oil has been a relatively cheap

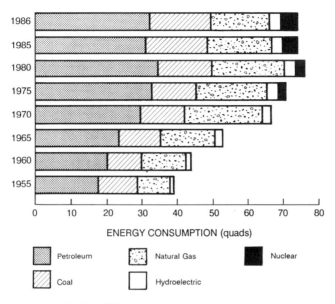

FIGURE 2   U.S. total energy consumption by source.

energy resource of widespread availability. Oil has a high energy density, which provides good vehicle driving range (see Figure 1). In addition, significantly improved combustion properties of petroleum-based fuels resulted from the advancement of fuel engine technology. The handling, storage, and use of petroleum-based fuels also have been optimized to the point that these procedures are both convenient and relatively safe. The internal combustion engine became the standard because of its inherent energy efficiency.

DEVELOPMENT OF EMISSION CONTROL FOR CONVENTIONAL VEHICLES

Because motor vehicles burn hydrocarbons using air as the source of oxygen, they emit a variety of combustion products, including carbon dioxide (CO_2), carbon monoxide (CO), unburned hydrocarbons (denoted HC), oxides of nitrogen (mainly NO with small amounts of NO_2, denoted NO_x), and other possible trace gases. Further, HC emissions can originate from fuel refining, transportation, and distribution. The emission of some of these gases (e.g., H_2O) has been considered to be of no consequence, while others (e.g., CO) have been recognized as requiring control when concentrations exceed public health-based air quality standards. Until recently the production of CO_2 by fossil fuel combustion was thought to

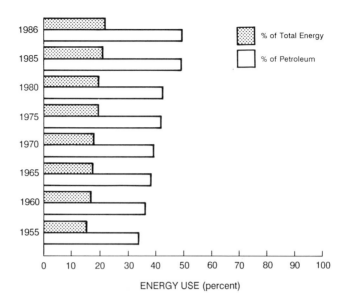

FIGURE 3 Percent of petroleum and total energy use by U.S. highway vehicles.

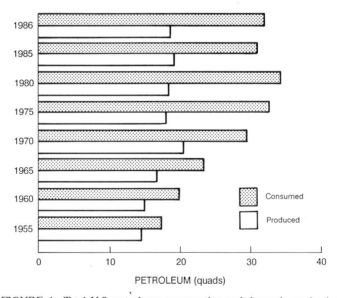

FIGURE 4 Total U.S. petroleum consumption and domestic production.

be environmentally innocuous, but fossil-based energy use from all sources over many decades has noticeably elevated the global concentration of this gas and raised the issue of the greenhouse effect (Helm and Schneider, in this volume; Schelling, in this volume). As vehicle use and density have increased in recent decades, the concentration of motor vehicle combustion gases increased to the point that the need for emission controls has been recognized and corrective action has been taken.

Ozone Formation

Before discussing ozone control strategies, it is important to understand the complexity of ozone generation. Ozone (O_3) is not emitted directly as a pollutant. Rather, it is formed by a variety of photochemical reactions among precursor pollutants as they mix in the troposphere and are irradiated by sunlight. In the troposphere, ozone is formed through the dissociation of NO_2 by sunlight to yield an oxygen atom, which then reacts with molecular oxygen (O_2) to form an O_3 molecule. If it is present, NO can react rapidly with O_3 to form NO_2 and an O_2 molecule. A steady-state, or equilibrium, concentration is soon established between O_3, NO_2, and NO, which in the absence of competing reactions exhibits low O_3 levels, which are of no concern. The injection of HC into the atmosphere upsets this equilibrium and allows ozone to accumulate at higher than steady-state concentrations by offering an alternate route for NO oxidation to NO_2. NO is emitted from high-temperature combustion sources and is converted to NO_2 in the atmosphere. The presence of HC, NO_x, and sunlight does not mean that the photochemical reactions will continue indefinitely. Terminal reactions gradually remove NO_2 and HC from the reaction mixtures, such that the photochemical cycles slowly come to an end unless fresh NO and HC emissons are injected into the atmosphere.

The Development of Emission Controls

From the early 1950s when Arie J. Haagen-Smit (1952) discovered the role of emissions in smog formation, a basis was developed upon which many steps have been taken to improve air quality. Haagen-Smit also observed that different organic materials give rise to different levels of photochemical smog, of which ozone is a principal component. These observations were verified in numerous smog chamber experiments, which eventually led the Los Angeles Air Pollution Control District to promulgate Rule 66 (Brunelle et al., 1966), thus making California the first state to legislate emission controls based on the reactivity of hydrocarbons to form smog. The emphasis of Rule 66 was to limit the emissions of highly reactive organic compounds more severely than emissions having relatively low reactivity. Rule 66 was the first effort to apply a practical hydrocarbon reactivity scale to a smog control program.

Among the first of the emission control measures taken by the automobile industry was the installation of crankcase blowby control devices to eliminate the emission of hydrocarbons from the crankcase by circulating them back into the engine to be burned. After the introduction of this system in California in 1961, it was incorporated into all the cars distributed nationwide in 1963. From 1968 through model year 1974, tailpipe exhaust emission control was implemented through engine calibration and modification to achieve more complete combustion along with the enhancement of desirable emission-reducing thermal reactions in the exhaust system. The trend in engine design up to model year 1970 had been toward engines with high compression ratios that required higher octane gasoline, but in 1971, anticipating the introduction of catalytic converters to reduce exhaust emissions further, manufacturers began to phase in the hardened valve seats and lower compression ratios needed for engines to burn the unleaded fuel required for proper operation of catalysts (Faith and Atkisson, 1972). In the 1975 model year, catalytic converters were introduced on motor vehicles, and today more than 80 percent of the in-use passenger car fleet has them (Motor Vehicle Manufacturers Association, 1987).

Regulation and control of hydrocarbon evaporative emissions from motor vehicle carburetors and fuel tanks were first applied in California to 1970 model year vehicles, and federal regulation followed in the next model year. In addition to vehicle evaporative control measures, refueling emission control equipment has been installed at motor fuel service stations in California, Washington, D.C., and parts of a few other states. Motor vehicle inspection and maintenance programs exist in more than half of the states for the identification and repair of vehicles with excessively high emissions. Transportation control measures, vehicle surveillance and recall programs, and environmental impact planning and permitting requirements also have been used to limit emissions.

Some of the more recent technological developments that aid in complying with the most stringent emission control requirements include computerized engine control, three-way catalysts (for HC, CO, and NO_x control), fuel injection systems (including optimized sequential injection), diagnostics (quicker and better repair reduces emissions), airflow mass sensors, platinum-tip spark plugs, and distributorless ignition.

Current passenger car exhaust emission standards require at least a 96 percent reduction in the emission of HC and CO and a 76 percent reduction in NO_x emissions compared with precontrol levels, as shown by the first three bars to the left in Figure 5. Light trucks (next three bars) must achieve a 90 percent reduction in the emission of HC and CO and a 67 percent reduction in NO_x emissions. Heavy gasoline trucks must also meet

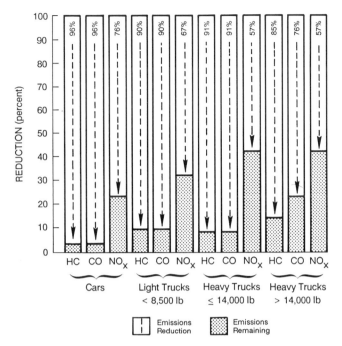

FIGURE 5 Mandated exhaust emission control reductions from precontrol levels (1991 gasoline motor vehicles).

stringent emission requirements, although there is more uncertainty in the exact magnitude of their reductions because of fewer available precontrol measurements.

ENVIRONMENTAL RESPONSE

Ambient Concentration Data

Success of pollution abatement activities for both mobile and stationary emission sources is best indicated by the response of the environment through the observed trends in air quality. Based on data from the U.S. Environmental Protection Agency (U.S. EPA, 1988a), the top half of Table 1 lists measured trend improvement in the ambient concentration of several atmospheric pollutants. The bottom part of Table 1 lists the results of measurements taken in the Lincoln Tunnel in New York City (Lonneman et al., 1986). Because the atmospheric pollution added to the atmosphere inside the Lincoln Tunnel is exclusively from motor vehicles, these data indicate that progress has been made in reducing emissions from motor vehicles. In contrast, with the exception of lead, the EPA ambient data

TABLE 1 Measured Air Quality Improvements

Pollutant (reference)	Base Year	Improvement Interval (years)	Improvement (percent)
Ambient data			
Lead (U.S. EPA, 1988a)	1977	9	87
Carbon monoxide (U.S. EPA, 1988a)	1977	9	32
Nitrogen dioxide (U.S. EPA, 1988a)	1977	9	14
Ozone (U.S. EPA, 1988a)	1979	7	13
Lincoln Tunnel air quality data			
Hydrocarbons other than methane (Lonneman et al., 1986)	1970	12	74
Carbon monoxide (Lonneman et al., 1986)	1970	12	76
Oxides of nitrogen (Lonneman et al., 1986)	1970	12	62

measures the aggregate trend in reducing emissions from all sources, which makes it impossible to assess the progress made by motor vehicles alone. Lead is an exception because its emission is almost exclusively due to the combustion of leaded gasoline, and therefore the measurement is not affected much by other sources. Lead reductions are seen to be the largest.

According to EPA air quality data for 1985–1987 (U.S. EPA, 1988b), a total of 68 areas were reported to have exceeded the ozone air quality standard in at least 1 year of the 3-year period, 6 more areas than previously reported. A total of 62 areas (11 in California) exceeded the ozone air quality standard in the 1984–1986 period (U.S. EPA, 1987c), 14 fewer areas than EPA's earlier compilation of 76 areas for the period 1983–1985. For many areas where ozone is measured, the summer of 1987 was hotter and more conducive to ozone formation than the previous period and is believed to be mainly responsible for the reported increase. A similar statement can be made for 1988. On the other hand, ambient levels of CO continued to decrease as expected, leaving 59 reported areas above the standard for the 2-year period 1986–1987, 6 fewer areas than in the previous 2-year period. Moreover, 23 of the 59 areas were in compliance with the CO air quality standard in 1987. Thus, even with the important progress achieved, a number of areas of the country still did not attain all air quality standards by the August 1988 congressional deadline.

National Ambient Air Quality Standards

The EPA currently uses 3-year time base data to determine if an area meets the air quality standard for ozone of 0.12 part per million (ppm). That is, the ozone standard is attained "when the expected number of days per calendar year with maximum hourly concentrations above 0.12 part per million (235 $\mu g/m^3$) is equal to or less than 1" (40 CFR 50.9). In general, most ozone monitor readings are below the standard. For example, Houston, considered by some to be the second worst area of the country for ozone pollution, had ozone levels below the standard for 99.47 percent of the hours monitored between 1981 and 1985 (American Petroleum Institute, 1987). The EPA uses the fourth highest daily maximum 1-hour average ozone value per three consecutive years of data to determine compliance with the standard, because the standard allows for an average of one incident above 0.12 ppm each year. The National Ambient Air Quality Standard (NAAQS) for CO is 9 ppm for an 8-hour nonoverlapping average and 35 ppm for a 1-hour average, neither of which is to be exceeded more than once a year. The 9 ppm 8-hour CO standard is the more stringent of the two and is the controlling standard. The nitrogen dioxide NAAQS is based on an annual mean of 100 $\mu g/m^3$ (0.053 ppm), not to be exceeded.

It is important to note that both emissions and ambient concentrations are subject to considerable variability. For example, the rate of automobile CO emissions has a characteristic pattern in time for most areas; it is highest during rush hours and lowest in the early morning. However, ambient CO concentrations can be highest during periods of lowest emissions if a very low inversion exists (low ambient mixing height), thus limiting dilution. As the sun rises, the mixing height increases. An additional source of variability exists for ozone because it is not emitted directly as a pollutant. As described above, ozone is formed by photochemical reactions of precursor pollutants as they mix in the troposphere. Thus, the concentration of ozone in time and space is a complex function of precursor emissions, weather, topography, and demographics.

Location and Severity of Violations

The Los Angeles area, where the air quality standard for ozone is exceeded on more than 140 days a year (for at least 1 hour per day), leads the nation in the frequency of such violations. The next most frequent occurrences of the standard being exceeded, all less than on 40 days a year, were recorded in a few metropolitan areas in Texas and the metropolitan areas along the northeast Atlantic coast (New York City and surrounding areas). Los Angeles is also the only area in the nation exceeding the air quality standard for NO_2. A combination of poor atmospheric ventilation, frequent sunshine, and bordering mountain barriers increases the potential for higher pollution levels in the Los Angeles area.

Forty years of regulatory effort and research and development resulted in tough emission control programs for both stationary and mobile sources. These efforts produced significant, but insufficient, air quality gains toward attaining established, but difficult, goals. Because of the special air quality situation in Los Angeles and the difficulty experienced in achieving air quality standards, the potential environmental benefits of both advanced conventional vehicles as well as alternatively fueled vehicles are being actively studied and will be described below.

ALTERNATIVE FUELS AND THEIR EFFECTS ON AIR QUALITY

Atmospheric emissions from alternatively fueled vehicles typically have a different composition from those produced by gasoline vehicles, which results in different reactivities for ozone generation as well. Many countries are experimenting with alternative power sources for transportation, as shown in Figure 6. Largely as a result of special circumstances (including heavy national subsidies in certain countries), the number of alternatively powered vehicles of all types may be as many as 7 million worldwide (U.S. Department of Energy, 1988). Countries that have achieved more than 10 percent use of alternative fuels for transportation are Brazil (ethanol, 23 percent) and the Netherlands (liquefied petroleum gas, 11 percent). Although this represents a small percentage (about 1 percent) of total world motor vehicles, much is being learned about the effects of alternative fuel strategies on air quality. The study of air quality effects due to the use of alternative fuels, as well as conventional fuels, has been based on several methods, including atmospheric modeling and smog chamber data using emission rates obtained from vehicles in use, prototype vehicles, and engineering judgment. The results from several studies suggest that under certain conditions, conversion of light-duty vehicles to methanol fuel has some potential to reduce ozone levels (Carter et al., 1986; Chang et al., 1989; Nichols and Norbeck, 1985; O'Toole et al., 1983a,b; Pefley et al., 1984; U.S. EPA, 1985, 1988c; Whitten and Hogo, 1983; Whitten et al., 1986). Much work needs to be done to determine the validity of the assumptions that underlie these conclusions and to understand better vehicle emission performance.

The EPA has issued some preliminary estimates of air quality benefit from the use of alternative fuels based on expected emissions and their associated ozone generation potential (U.S. EPA, 1987b). The alternatives

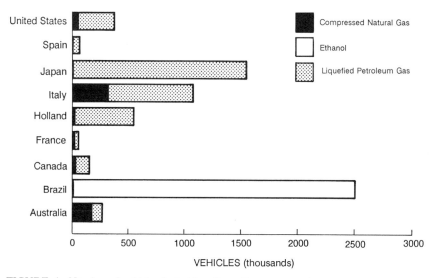

FIGURE 6 Number of vehicles fueled by alternative energy systems.

addressed include gasoline blends with 10 percent ethanol (gasohol), 5 percent methanol, and 11 percent methyl tertiary butyl ether (MTBE), as well as compressed natural gas (CNG) and various methanol fuels (100 percent methanol, referred to as M100, and a mixture of 85 percent methanol and 15 percent gasoline, known as M85).

Because gasoline blends contain more oxygen than gasoline alone, they contribute to a leaner combustion process for vehicles with limited or no self-compensating ability to adjust to changes in fuel oxygen. Obviously, leaner combustion also could be accomplished with engine calibration. Therefore, blends are a method to lean-out the vehicle fleet after the vehicles are built. Blends might be used at high-altitude locations where the vehicle fleet may be running richer than at low-altitude locations. Thus, under certain circumstances, the use of blends leads to lower emissions of hydrocarbons (1–9 percent) and carbon monoxide (10–30 percent) at the expense of some increase in NO_x emissions (4–6 percent). Based on this fact, Colorado started a special blend program to reduce wintertime carbon monoxide levels in the mile-high city of Denver in December 1987 (Miron et al., 1986; Regulation No. 13, 1987) and is in the process of evaluating its effectiveness. However, blends are much less effective for the newer-technology vehicles having adaptive fuel metering systems with nonvolatile computer memory and exhaust gas oxygen feedback control. These systems automatically adjust to the amount of oxygen in the exhaust gas in order to achieve optimal performance from the three-way catalyst. In the future, blends are likely to be used to reduce emissions only in special situations as long as they continue to be effective (depending on the number of older vehicles still in service).

Vehicles operating on compressed natural gas, and emitting mostly methane, would also be expected to have low ozone generation potential because methane has very low reactivity and generates essentially no ozone. However, CNG vehicles do produce some reactive hydrocarbons during the combustion process and therefore emit some hydrocarbons other than methane. In addition, these vehicles can have higher NO_x emissions because of the relatively high flame temperature (U.S. EPA, 1987a).

Methanol as a Motor Fuel

The potential air quality benefits of neat (M100) or near neat (M85) methanol are related to the fact that the organic emissions from vehicles designed to operate on methanol primarily consist of unburned methanol, which has significantly lower photochemical reactivity for ozone formation than typical hydrocarbon emissions from gasoline-fueled vehicles. Two requirements are necessary before a beneficial effect on air quality can be realized: (1) emissions of other coemitted substances must not significantly offset the lower reactivity of primary emissions and (2) the total mass of emissions compared with conventional vehicles should not be high enough to offset the benefit of lower reactivity. Methanol-fueled vehicles emit formaldehyde, produced by the partial oxidation of methanol, as well as nonoxygenated hydrocarbons that are more chemically reactive in the atmosphere than methanol. Catalysts can be used to remove most of the formaldehyde before it is emitted. However, vehicle data suggest that it does not appear possible, with current technology, to control continuously the more reactive formaldehyde emissions from methanol vehicles to the low level typical of gasoline vehicles over their expected lifetime (Nichols et al., 1988). It is simply expecting too much for a catalyst to maintain nearly 100 percent efficiency over vehicle lifetime. Another issue of particular importance is the failure of catalysts and the occasions on which catalysts are removed, because formaldehyde is both an eye and nose irritant. It is also not yet clear what the total mass of emissions will be under typical in-use conditions during the life of a methanol vehicle. More developmental work is essential to understand better the potential of methanol fuel and to define vehicle emission limits accurately.

The recommended methanol fuel specification of Ford Motor Company is M85. Several safety-related considerations prompt this specification: the 15 percent gasoline provides flame luminosity (a pure methanol flame is invisible in daylight); it makes the vapor above the liquid in the fuel tank too rich to ignite at normal operating temperatures; and it acts as

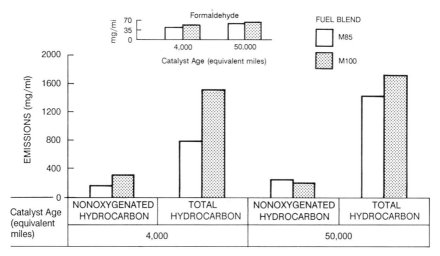

FIGURE 7 Organic emissions, from a flexibly fueled Ford Escort using M85 and M100, as a function of catalyst age. Even though M100 contains no nonoxygenated hydrocarbons, its emissions are comparable with M85.

an odorant that discourages ingestion. In addition, the 15 percent gasoline provides the vapor pressure necessary to cold start an engine at low ambient temperatures and may improve engine durability. Theoretically, M100 should be a better fuel than M85 from an environmental standpoint because M100 does not contain any nonoxygenated hydrocarbons to emit. However, preliminary emissions testing of current-technology flexible-fuel vehicles equipped with laboratory-aged exhaust catalysts (simulating 4,000 and 50,000 miles of driving) indicates somewhat higher emissions of formaldehyde and methanol with M100 than with M85. In addition, as shown in Figure 7, nonoxygenated hydrocarbon emissions with M100 fuel were comparable with those of M85 (Nichols et al., 1988), even though M100 contains no nonoxygenated hydrocarbons. This is because complicating factors such as hydrocarbon formation in the combustion chamber from fuel and oil (with both M85 and M100) and harder starting with M100, cause nonoxygenated hydrocarbons to be emitted out of proportion to their presence in the fuel (Gold and Moulis, 1988), at least for vehicles using current technology.

Thus, it is not clear yet whether technological advances will make one fuel specification better than the other from an environmental standpoint without compromising important vehicle operating and safety requirements. For example, in one advanced-concept methanol prototype vehicle, a Toyota lean-burn system with swirl control, the average low mileage exhaust hydrocarbons and formaldehyde were nearly equivalent with M85 and M100

(Murrell and Piotrowski, 1987). However, higher hydrocarbon levels in some tests with M100 may have resulted from difficulties experienced in starting. Evaporative emissions, although below gasoline vehicle evaporative standards, were higher on M85 than M100. Until additional data indicate the contrary, M85 and M100 can be considered nearly equivalent from an environmental standpoint. In terms of other important aspects listed earlier, M85 is preferred.

Computer Simulation Studies (Single Day)

Over the past few years, numerous computer simulation studies of atmospheric chemistry have been undertaken to model the impact of motor vehicle methanol fuel substitution (e.g., Chang et al., 1989; Nichols and Norbeck, 1985; O'Toole et al., 1983a,b; Pefley et al., 1984; Whitten and Hogo, 1983; Whitten et al., 1986). Results from these studies are summarized in Table 2. These simulations were based mainly on "trajectory models," such as the city-specific empirical kinetic modeling approach (U.S. EPA, 1977), and thus share the characteristic that a time-dependent emissions profile, called a trajectory, is specified as input.

A typical simulation might be undertaken to seek the upper limit for ozone reduction, assuming that gasoline vehicles have been completely replaced by methanol vehicles. Because actual motor fleet turnover (i.e., new vehicles completely replacing old) would occur over some characteristic period of time (up to several decades), such "all-methanol" scenarios are somewhat academic. Details on the substitution assumptions are listed in Table 2. Another important input parameter that must be specified for a model is the total mass of the emissions. As noted, this quantity is not well understood, and therefore an equal-carbon basis is usually assumed. That is, because a methanol molecule contains one carbon atom, a number of methanol molecules equal to the number of carbon atoms contained in the average exhaust and evaporative molecule is used (see Figure 8). Therefore, an all-methanol trajectory would be one in which each vehicle releases an amount of methanol exhaust and evaporative mixture that contains the same number of carbon atoms as would be emitted if gasoline vehicles were used instead. In so doing, the total mass of emitted carbon presumably is representative of current transportation load, and the effects of the differing exhaust chemicals can be assessed. Three all-methanol simulations are listed in Table 2.

One-day modeling simulations are highly sensitive to input variables such as the ratio of reactive hydrocarbons to NO_x, ozone concentration aloft, initial pollutant concentrations, and hydrocarbon and NO_x emissions along the trajectory. Because of these sensitivities, and because of the

TABLE 2 Single-Day Modeling Studies

Group, Sponsor (reference)	City Modeled, Year (vehicle substituted)	Emissions Relative to Gasoline (percent)[a]					Study Notes	Peak Ozone Reduction (percent)
		ME	F	MN	HC	NO_x		
Systems Applications, Inc., ARCO (Whitten and Hogo, 1983)	Los Angeles, 1987 (all gasoline vehicles)	100	0	0	0	100	b	31
		90	10	0	0	100	b	22
		89	10	1	0	100	b	19
		80	20	0	0	100	b	13
University of Santa Clara, Department of Energy (Pefley et al., 1984)	Los Angeles, 1987 (all gasoline vehicles)	90	10	0	0	100	b	18
Jet Propulsion Laboratory, California Institute of Technology, National Aeronautics and Space Administration (O'Toole et al., 1983a)	Los Angeles, 1982 (all gasoline vehicles)	0	0	0	0	0	c	25
		56	14	0	30	100		14
		56	14	0	30	50		17
		28	7	0	15	50		20
Ford Motor Company (Nichols and Norbeck, 1985)	20 cities, 1982 (light-duty cars, trucks)	100	0	0	0	100	b	13
		90	10	0	0	100	b	3
Systems Applications, Inc., Environmental Protection Agency (Whitten et al., 1986)	Philadelphia, 2000 (all motor vehicles)	0	0	0	0	0	c	1
		100	0	0	0	100	b	18
		110	6	0.6	0	100	b	4
		105	12	0.6	0	100	b	2

Table 2 continues on the following page

TABLE 2 Continued

Group, Sponsor (reference)	City Modeled, Year (vehicle substituted)	Emissions Relative to Gasoline (percent)[a]					Study Notes	Peak Ozone Reduction (percent)
		ME	F	MN	HC	NO_x		
Ford Motor Company (Chang et al., 1989)	20 cities 1985 (light-duty cars, trucks)	53	3.5	0	43.5	100		3
		86	1.5	0	12.5	100		5
		90	10	0	0	100		4
		0	0	0	0	100	b	17
		53	3.5	0	43.5	100		1.3
	2000	86	1.5	0	12.5	100		2.6
		90	10	0	0	100		1.9
		0	0	0	0	100	b	7.0

NOTE: The all-methanol scenarios are academic and suggest only the upper limit for ozone reduction at equal carbon and NO_x levels.

[a] Composition expressed as the percentage of methanol (ME), formaldehyde (F), methyl nitrite (MN), and nonoxygenated hydrocarbons of vehicle-related emissions compared with the conventional emissions they replace. The percentage refers to equivalent emissions of carbon, excluding carbon dioxide. Scenarios that do not add up to 100 percent indicate that more or less of the baseline gasoline carbon emissions were substituted in the computer simulation.
[b] Nonoxygenated hydrocarbon emission rates were assumed to be zero for methanol vehicles in these scenarios. Methanol vehicles, however, do emit some nonoxygenated hydrocarbons (see Figure 7).
[c] No vehicles.

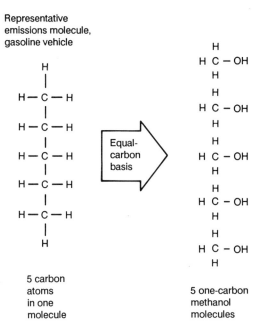

FIGURE 8 Equivalent carbon-for-carbon basis for one typical emission molecule from a gasoline vehicle. Note that both evaporative and exhaust emissions are composed of many different types of molecules; typically, equal-carbon basis calculations are carried out for each chemical species.

modeling assumptions discussed here and listed in Table 2, one must be cautious when interpreting simulation results.

The results of trajectory model simulations such as those found in Table 2 have been used to estimate the amount of ozone produced relative to gasoline hydrocarbon emissions, assuming an equal-carbon emission basis (Chang et al., 1989; EPA, 1987a; Gold and Moulis, 1988; Shiller, 1987). For most cases, the effect of a switch to methanol, including the effect of emitting some formaldehyde, was investigated under the assumption that the NO_x emission rate from methanol vehicles is the same as that from gasoline vehicles. The most recent calculated average reactivity factors for 20 cities were about 0.58 for methanol and 2.15 for formaldehyde, compared with 1.0 for conventional hydrocarbon ozone generation potential at equal carbon emissions. These factors suggest that formaldehyde (carbon atom for carbon atom) is more than twice as efficient as hydrocarbons in generating ozone, whereas methanol is less than about 60 percent as efficient as hydrocarbons. Previously, the difference between the calculated reactivity factors for methanol and formaldehyde were larger (U.S. EPA,

1986a, 0.02 and 2.95; Nichols and Norbeck, 1985, 0.38 and 4.8; and U.S. EPA, 1987a, 0.43 and 4.83).

For a few of these simulations, reductions in NO_x emissions were also evaluated, which caused either an increase or a decrease in projected ozone levels. This, for example, is illustrated by comparing different impacts of NO_x reductions on ozone projections for Los Angeles (cases 2 and 3 of the Jet Propulsion Laboratory study in Table 2) and Philadelphia (cases 1 and 2). The Philadelphia study suggests that NO_x control can be counterproductive to ozone control, whereas results for Los Angeles suggest that the opposite can be true. In general, the greater the impact of NO_x control on reducing ozone (greater impact with increasing ambient HC/NO_x ratio), the less important reactivity differences become (Dodge, 1984) and the smaller is the projected impact of methanol-fueled vehicles. Consistent with this interpretation, the Philadelphia simulation seems to be more sensitive to reactivity changes than the Los Angeles simulation. Of course, projected results depend on the accuracy of the chemical mechanism and emission inventory assumptions used in the model (which determines the local HC/NO_x ratio) and the particular wind flow field selected for the simulation. Further study is necessary to determine how general the results are for each city and their implications for ozone control planning. In addition, because ozone builds up more slowly with methanol, it is necessary to determine the impact of methanol vehicles under multiday episodes of stagnant air, which provide more time for ozone formation.

Smog chamber studies of simulated urban air (Carter et al., 1986) confirm the results of single-day simulations that lower peak ozone levels occur for the first day for the case in which one-third of the total vehicle emissions are from methanol-fueled engines. However, the ozone levels in the chamber rose over the course of the second and third days, becoming similar to those observed in the control case of 100 percent gasoline engine emissions. This suggests that even if one-third of the emissions from gasoline vehicles is replaced with methanol emissions, little or no benefit will be achieved for multiday ozone episodes. Because of the computer model sensitivities noted above, and because smog chamber simulations may be affected by the greater surface-to-volume effects inherent in a chamber, further study is needed to strengthen these findings. At this stage, it is prudent to interpret these results qualitatively, rather than quantitatively, when discussing this matter.

Multiday Simulation Studies

The scope of ozone modeling is being extended to study the impact of methanol-fueled vehicles during multiday episodes. The most noteworthy multiday study, employing a Cray X-MP/48 supercomputer, was conducted at Carnegie Mellon University under the sponsorship of the California Air Resources Board (CARB). Some of the results have been released (California Air Resources Board, 1988; *Computers in Science*, 1988; Harris et al., 1988). A sequential modeling approach was used to guide the study effort, starting with chemical kinetic modeling, then moving to trajectory analysis, and finishing with airshed modeling of the entire California South Coast Air Basin. For the final airshed modeling analyses, 13 control strategy scenarios reflecting projected emission assumptions and expected conditions in the basin for the year 2000 have been run.

Vehicle emission rate assumptions were supplied by CARB (1988) for six cases (three with conventional fuels and three with methanol). This matrix is shown in Table 3. Emission rates for the first case of conventionally fueled vehicles were based on CARB's EMFAC7C data (CARB, 1986) and reflect all post-1987 regulations that have been adopted or implemented. The second case reflects additional conventional control measures that are being considered for adoption in the 1990s (e.g., 0.25 g/mi nonmethane hydrocarbons [NMHC] beginning in 1995 and 0.2 g/mi NO_x beginning in 1997) if they can be shown to be feasible and desirable. This was simulated by adjusting the EMFAC7C data by the ratio of proposed standards to existing standards. The third case for advanced conventional vehicles assumes that all motor vehicles will be designed to meet the lowest foreseeable emission standards and will perform at the lowest end of the in-use emission range that might be possible with further technical development.

The next three cases are methanol scenarios. Each assumes that 100 percent of the 1990–2000 model year highway vehicles are built to run on methanol, with a similar percent conversion for off-road mobile sources and retail gasoline distribution. The scenario labeled case four explores the effect of using M85 by assuming a 15 mg/mi emission rate of formaldehyde for cars and light trucks; a contribution for nonoxygenated hydrocarbon emissions was also included. Case 5 assumes the same (case 2) vehicle fueled with M100, carbon for carbon, a comparatively high formaldehyde emission rate of 55 mg/mi, and zero nonoxygenated hydrocarbon emissions. Case 6 is identical to case 5 except that a lower value for formaldehyde emission, 15 mg/mi, is assumed. Note that both of these M100 cases assume no nonoxygenated hydrocarbon emissions, which is contradictory to the M100 data reported in Figure 7. In general, some of these emission rate assumptions are optimistic and have not been shown to be feasible; the results of these simulations appear in Table 4. Note that a seventh case also appears in this table.

Case 1 was selected as the base case by which we can compare the reductions predicted by the other simulations. Comparisons are reported in terms of the percent of the unachieved ozone reduction required to bring

TABLE 3 Multiday Study Simulation Matrix

		FUEL TYPE	
		Gasoline	Methanol
V E H I C L E T Y P E		Case 1: Current conventional vehicles (normal fleet turnover)	
		Case 2: 1990–2000 model year vehicles (additional regulations, normal fleet turnover to 2000)	Case 4: M85 @ 100 percent substitution[a] 15 mg/mi formaldehyde, ? HC Case 5: M100 @ 100 percent substitution[a] 55 mg/mi formaldehyde, no HC Case 6: M100 @ 100 percent substitution[a] 15 mg/mi formaldehyde, no HC
		Case 3: Advanced vehicles (technological break- through) and 100 percent substitution	

[a]Complete replacement of methanol vehicles for both highway and off-road mobile vehicles and replacement of methanol for gasoline in the retail fuel distribution system.

the base case into compliance (100 percent) and appear in Figure 9. For the base case scenario, the maximum peak ozone level was calculated to be 0.27 ppm. Because 0.12 ppm is the value specified by the ozone standard, a 57 percent reduction in ozone level would be required to achieve compliance; this is represented by the top bar of Figure 9 plotted at 100 percent and 0.27 ppm. From Table 4, we note that cases 2 and 3 were found to reduce ozone levels by 3 and 9 percent, respectively; this corresponds, respectively, to 5 and 16 percent of the reduction required to reach the 0.12 ppm level (greater by the ratio of 100/57). These two cases are indicated by bars 2 and 3 of Figure 9. The next three bars indicate percent of achievement for the three methanol cases. Note that the best methanol case, M100 with low formaldehyde, achieves only 29 percent of the reduction required to reach 0.12 ppm. These preliminary results of this study are consistent with previous research; however, the reader is again reminded that these

TABLE 4 Multiday Modeling Studies

Group, Sponsor (reference)	City Modeled, Year (vehicle substituted)	Emissions Relative to Gasoline[a] (percent)					Study Notes	Peak Ozone Reduction (percent)
		ME	F	MN	HC	NO_x		
Carnegie Mellon University, CARB (CARB, 1988)	Los Angeles 2000 (all vehicles)	Some details not available						
	Case 1	Conventional vehicle base case						0
	2	1900–2000 model year vehicles						3
	3	Most advanced conventional vehicles						9
	4	M85	Low F	?	?	?	d	9
	5	M100	High F	?	0	?	b	10
	6	M100	Low F	?	0	?	b	16
	7	0	0	0	0	0	c	24

[a] Composition expressed as the percentage of methanol (ME), formaldehyde (F), methyl nitrite (MN), and nonoxygenated hydrocarbons (HC) of vehicle-related emissions compared with the conventional emissions they replace. The percentage refers to equivalent emissions of carbon, excluding carbon dioxide. Methanol vehicles, however, do emit some nonoxygenated hydrocarbon emission rates were assumed to be zero for methanol vehicles in these scenarios. Methanol vehicles, however, do emit some nonoxygenated hydrocarbons (see Figure 7).
[b] Nonoxygenated hydrocarbon emission rates were assumed to be zero for methanol vehicles in these scenarios.
[c] No vehicles.
[d] With nonoxygenated hydrocarbons.

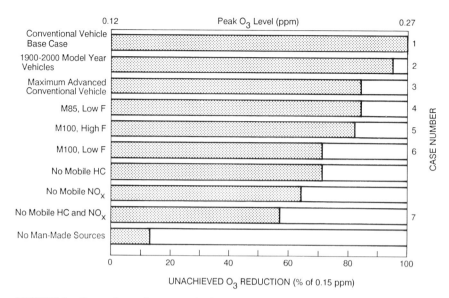

FIGURE 9 Comparison of ozone reduction strategies for the South Coast Air Basin for the year 2000. The solid bars show the percent of unachieved reduction of ozone levels from a base case value of 0.27 ppm to the 0.12 ppm level required by the ozone standard. Case numbers are the same as in Tables 3 and 4 and are described further in the text. These results suggest that no strategy limited to the motor vehicle transportation system alone will be sufficient to achieve the ozone standard in the Los Angeles area. SOURCE: California Air Resources Board, 1988.

findings depend on the plausibility of the emission rate assumptions and initial conditions.

The bars below case 6 in Figure 9 show the reductions achieved by eliminating all ozone precursor emissions due to the motor vehicle transportation system, that is, emissions from all retail gasoline and diesel distribution centers, all petroleum refining operations, and all motor vehicles (both highway and off-road vehicles) except motorcycles. The three cases studied were as follows: the removal of all HC with reduced NO_x achieved a reduction of 16 percent, the removal of all NO_x with reduced HC achieved 20 percent, and the removal of all HC and NO_x (case 7) achieved 24 percent, far short of the 57 percent required. Further, a peak ozone level of 0.139 ppm was estimated for the extreme case of removing all in-basin emissions (all sources) and assuming low-level initial conditions (low pollutant concentrations entering the basin). Even for this case, only 87 percent of the reduction required to achieve the ozone standard was attained, as shown by the bottom bar of the figure.

These results suggest that achievement of the ozone standard in the

Los Angeles area will not be possible with any strategy that targets mobile sources alone. It may not be possible even with the removal of all manmade sources.

The results further indicate that methanol-fueled vehicles do have some potential to reduce ozone in the Los Angeles area but that the magnitude of the reduction during multiday episodes is likely to be smaller than that of earlier single-day modeling results (see Table 2 for Los Angeles). The emission assumptions for the M85 case are considered to be more realistic (for either the M85 or the M100 case). This suggests a 6 percent ozone benefit (9 percent versus 3 percent, respectively) for methanol-fueled vehicles over conventionally fueled vehicles (with additional controls) may be possible if the assumed optimistic emission rates can be achieved. More work is needed to examine the accuracy of the emission assumptions and the general applicability of these results to other modeling conditions.

Public Health and Direct Formaldehyde Emissions

For any ozone control strategy, it is vital to consider the possibility of undesirable effects that may arise as a consequence of the control strategy itself. In other words, we must make sure that the "cure is not worse than the disease." Thus, it is necessary to evaluate the effect of direct emissions of formaldehyde on public health, particularly in confined spaces such as parking garages. Work is being done to gain a better understanding of this concern. For example, the Health Effects Institute (HEI) has undertaken a study of formaldehyde to complement their earlier study on the health effects of direct methanol emissions (HEI, 1985). The earlier study indicated that adverse human health effects due to methanol at projected exposure levels associated with a move to methanol fuel are unlikely. In another study (Gold and Moulis, 1988), EPA's proposed methanol vehicle emission standards (U.S. EPA, 1986b) are shown to prevent concentrations of methanol and formaldehyde from reaching toxic levels in most commonly encountered driving scenarios. However, further analysis of parking garages and other areas with restricted air dilution is needed before all concerns can be fully resolved. As for open and unrestricted spaces, the Carnegie Mellon study (*Computers in Science*, 1988) showed that 90 percent of atmospheric formaldehyde is created by photooxidation of hydrocarbons and that only 10 percent comes from direct emissions. Thus, a significant increase in average ambient formaldehyde levels with methanol vehicles is unlikely.

PUBLIC POLICY STRATEGIES

In recognition of the difficult task of achieving standards in the Los Angeles Basin, the South Coast Air Quality Management District (SCAQMD) and the Southern California Association of Governments (SCAG) have announced a three-tier strategy to attain air quality standards over the next 20 years (SCAQMD and SCAG, 1987, 1988). The plan, which was proposed in June 1988, acknowledges that full-scale implementation and advanced development of known technology will be inadequate to achieve the standards in Los Angeles. Significant technological breakthroughs are required.

To illustrate the magnitude of the situation, Figure 10 shows the 1985 emission inventory for both mobile and stationary sources of reactive hydrocarbons and oxides of nitrogen in the South Coast Air Basin (SCAQMD and SCAG, 1987). Based on estimates by SCAQMD and SCAG, a 79 percent reduction in reactive hydrocarbons from the 1985 inventory would be necessary to attain the air quality standard for ozone. Elimination of highway vehicle hydrocarbon emissions would only provide a 43 percent reduction, far short of the required 79 percent. Therefore, the demands of attaining the ozone standard in Los Angeles would exceed even the most aggressive application of current motor vehicle technology. Note that this conclusion is consistent with the multiday computer simulations described above.

The three-tier strategy proposed by SCAQ to reduce emissions to the point where all air quality standards are achieved by 2007–2010 requires a full-scale implementation of known technology (tier 1), significant advancement of known technology (tier 2), and technological breakthroughs (tier 3). In the second phase of the plan, it is assumed that 40 percent of passenger cars, 70 percent of freight vehicles, and 100 percent of buses will use clean-fuel technologies (Acurex, 1986). In January 1988, SCAQMD adopted a five-year $30-million Clean Fuels Program, which includes 19 mobile source related demonstration projects for technologies, such as electric vehicles, methanol, and compressed natural gas. Tier 3 assumes full electrification of all motor vehicles and stationary combustion sources. That assumption relies on new technologies such as superconductors and improved electrical storage devices and either the building of new infrastructures or the elimination of existing infrastructures or both. Tier 3 is clearly beyond current capability.

CUSTOMER VALUE AND FUEL STRATEGY

If a conversion to alternative fuels is to be successful, the vehicles that use these fuels must represent value to the customer or they will not sell.

Customer surveys, such as the 1987 Automotive Consumer Profile survey of 5,000 members of the driving-age public, show that reliability and quality lead the list of consumer "wants" (see Figure 11), with safety and service (availability of parts and labor) also high on the list (*Power Report*, 1988). About 70 percent or more of the respondents said that these features were "very or somewhat important" in the purchase of their next vehicles. Purchase price, operating cost, and resale value (a factor in operating cost) are also perceived as important. Based on new car buyers' perceptions of how these features would be affected by a switch to methanol fuel (see Figure 12), the perceived benefits of methanol are in areas of relatively low importance to consumers.

Vehicles that use alternative fuels must be fully competitive in value with current transportation alternatives. In fact, it could be argued that they must offer an advantage to overcome any concerns a customer might have. This advantage may have to take the form of buyer incentives (at least in the beginning) to induce consumer risk investment by compensating for the perceived potential disadvantages associated with the convenience, reputation, and resale of alternative-fuel vehicles in relation to gasoline-fueled vehicles. It is critical that consumer concerns be addressed at the outset.

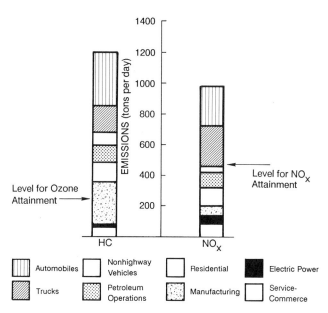

FIGURE 10 South Coast Air Basin emissions by source for 1985. In 1985 the ozone produced by precursor sources unrelated to transportation was sufficient to exceed the standard.

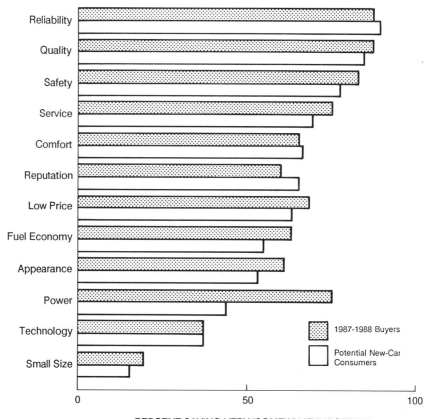

FIGURE 11 Important factors affecting new-car purchase decisions.

From the customer's point of view, the availability of vehicles capable of using alternative energy sources will alone not produce a change in the present transportation system. There must also be a substantial increase in the production capacity for the alternative fuel in question and a distribution network to make such fuel readily available. Therefore, the development of vehicles having "fuel flexibility" seems imperative in any program to introduce a new fuel.

CONCLUSIONS

Significant advances have been made in reducing emissions from current-technology vehicles. Further reductions are expected as newer vehicles replace old. Even with the progress to date, certain areas of the country still did not attain all air quality standards by the August 1988 congressional deadline. The Los Angeles area faces the most difficult task to achieve the national air quality standards.

Because of the special air quality situation in Los Angeles, the potential environmental benefits of alternative fuels are being studied to determine what can be accomplished. To date, the relatively low cost of gasoline and diesel fuel and their other inherent advantages have restricted the use of other fuel sources. The renewed interest in alcohol as a fuel in the United States results mainly from a hope that it may reduce environmentally undesirable emissions. Studies suggest that conversion of light-duty vehicles to methanol fuel has some potential to reduce ozone levels if emission assumptions are correct, but they also suggest that in certain areas it is not possible to achieve the ozone standard with any strategy that targets only the emissions from the motor vehicle transportation system. Much remains to be done to ascertain the validity of the assumptions that underlie the conclusions reached thus far and to gain a better understanding of currently

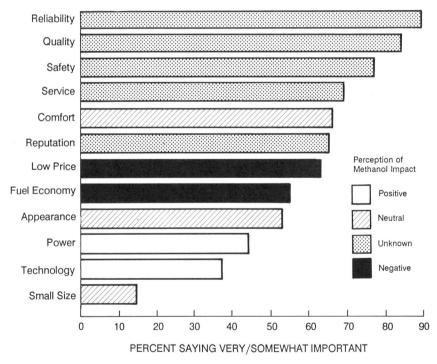

FIGURE 12 Consumer perceptions of the effect of methanol-fuel capability on factors important to a new-car purchase.

unregulated emissions. The improvements made in conventional technology also need careful evaluation for comparison.

Surveys of motor vehicle consumer preferences indicate that the perceived benefits of methanol are relatively less important than other vehicle attributes. This situation poses an even greater challenge for vehicle manufacturers and governmental regulators to deliver these new technology vehicles in a way that provides equal, if not greater, value to the customer.

REFERENCES

Acurex. 1986. California's Methanol Program, Evaluation Report, Vol. I, Executive Summary. November. Mountain View, Calif.: Acurex Corp.

American Petroleum Institute (API). 1987. Ozone Concentration Data Analysis. Washington, D.C.: API.

Bolt, J. A. 1980. A Survey of Alcohol as a Motor Fuel. SAE Technical Paper Series, SAE/pt-80/19. Warrendale, Pa.: Society of Automotive Engineers.

Brunelle, M. F., J. E. Dickinson, and W. J. Hamming. 1966. Effectiveness of Organic Solvents in Photochemical Smog Formation: Solvent Project, Final Report. Los Angeles County Air Pollution Control District.

California Air Resources Board (CARB). 1986. Methodology to Calculate Emission Factors for On-Road Motor Vehicles. CARB Technical Support Division, Emission Inventory Branch, Motor Vehicle Emissions and Projections Section. Sacramento, Calif.: CARB.

California Air Resources Board. 1988. Selected draft of Carnegie Mellon University study results released in response to a request for information in preparation for the June 1, 1988, CARB workshop on formaldehyde emission standards.

Carter, W. P. L., R. Atkinson, W. D. Long, L. L. N. Parker, and M. C. Dodd. 1986. Effects of Methanol Fuel Substitution on Multi-Day Air Pollution Episodes. University of California for California Air Resources Board, Contract No. A3-125-32. Sacramento, Calif.: CARB.

Chang, R. Y., S. J. Rudy, G. Kuntasal, and R. A. Gorse, Jr. 1989. Impact of methanol vehicles on ozone air quality. Atmospheric Environment (In Press).

Chemical Rubber Company. 1983. Handbook of Chemistry and Physics, 64th ed. Boca Raton, Fla.: CRC Press.

Computers in Science. 1988. Carnegie Mellon University. January–February.

Dodge, M. C. 1984. Combined effects of organic reactivity and $NMHC/NO_x$ ratio on photochemical oxidant formation—A modeling study. Atmospheric Environment 18(8):1657–1665.

Faith, W. L., and A. A. Atkisson, Jr. 1972. Air Pollution, 2d ed. New York: John Wiley & Sons.

Gold, M. D., and C. E. Moulis. 1988. Effects of emission standards on methanol vehicle-related ozone, formaldehyde, and methanol exposure. APCA Paper No. 88-41.4, SDSB/EPA. Pittsburgh, Pa.: Air Pollution Control Association.

Haagen-Smit, A. J. 1952. Chemistry and physiology of Los Angeles smog. Industrial and Engineering Chemistry 44(6):1342.

Hagen, D. L. 1977. Methanol as a fuel: A review with bibliography. SAE Technical Paper Series, SAE-770792. Warrendale, Pa.: Society of Automotive Engineers.

Harris, J. N., A. G. Russell, and J. B. Milford. 1988. Air Quality Implications for Methanol Fuel Utilization. SAE Technical Paper Series, SAE-881198. Warrendale, Pa.: Society of Automotive Engineers.

Health Effects Institute (HEI). 1985. Letter to Dr. Bernard Goldstein, Assistant Administrator for Research and Development, EPA, from Thomas P. Grumbley. August 12.

Ingersoll, E. P. 1895. The Horseless Age (November).

Ingersoll, E. P. 1897a. The oil famine bugaboo, a gunpowder motor. The Horseless Age II(3).
Ingersoll, E. P. 1897b. Acetylene as a motive agent, motor cabs (electric) in New York. The Horseless Age (February).
Ingersoll, E. P. 1897c. Acetylene motors. The Horseless Age (March).
Ingersoll, E. P., ed. 1897d. Riker electric victoria carbonic acid carriage motor, the Worthley steam carriage, new motor (electric) fire (fighting) apparatus, alcohol as a fuel for motors. The Horseless Age (September).
Ingersol, E. P. 1897e. Compressed air vehicles of the Pneumatic Carriage Company. The Horseless Age (October).
Lonneman, W. A., S. A. Meeks, and R. L. Sella. 1986. Non-methane organic composition in the Lincoln Tunnel. Environmental Science and Technology 20:790–796.
Miron, W. L., R. A. Ragazzi, T. W. Hollman, and G. L. Gallagher. 1986. Ethanol-blended fuel as a CO reduction strategy at high altitude. SAE Technical Paper Series, SAE-860530. Warrendale, Pa.: Society of Automotive Engineers.
Motor Vehicle Manufacturers Association (MVMA). 1987. MVMA Motor Vehicle Facts and Figures '87. Washington, D.C.: MVMA.
Murrell, J. D., and G. K. Piotrowski. 1987. Fuel economy and emissions of Toyota T-LCS-M methanol prototype vehicle. SAE Technical Paper Series, SAE-871090. Warrendale, Pa.: Society of Automotive Engineers.
Nichols, R. J. 1986. The flexible fuel vehicle: The bridge to methanol. Paper FL-86-103. National Petroleum Refiners Association, Fuels and Lubricants Meeting, Houston, Texas. November.
Nichols, R. J., and J. M. Norbeck. 1985. Assessment of emissions from methanol-fueled vehicles: Implications for ozone air quality. APCA Paper No. 85-38.3. Pittsburgh, Pa.: Air Pollution Control Association.
Nichols, R. J., E. Clinton, E. T. King, C. S. Smith, and R. J. Wineland. 1988. A view of FFV aldehyde emissions. SAE Future Transportation Technology Conference and Exposition, August 8–11.
O'Toole, R., E. Dutzi, R. Gershman, W. Heft, W. Kalema, and D. Maynard. 1983a. California Methanol Assessment, Vol. 1: Summary Report: Jet Propulsion Laboratory and California Institute of Technology, N83-33340. Springfield, Va.: National Technical Information Service.
O'Toole, R., E. Dutzi, R. Gershman, W. Heft, W. Kalema, and D. Maynard. 1983b. California Methanol Assessment, Vol. 2: Technical Report: Jet Propulsion Laboratory and California Institute of Technology, N83-33341. Springfield, Va.: National Technical Information Service.
Pefley, R. K., B. Pullman, and G. Whitten. 1984. The impact of alcohol fuels on urban air pollution: Methanol photochemistry study. University of Santa Clara and Systems Applications, Inc., for Department of Energy, DOE/CE/50036-1. November. Springfield, Va.: National Technical Information Service.
Power Report. 1988. Trouble-free, quality, safe cars important to consumers. The Power Report Newsletter (March).
Regulation No. 13. 1987. The reduction of carbon monoxide emissions from gasoline powered motor vehicles through the use of oxygenated fuels. Adopted by the Colorado Air Quality Control Commission on July 29, 1987, and submitted to the Colorado Secretary of State for publication in the Colorado Register on July 10, 1987.
South Coast Air Quality Management District (SCAQMD) and Southern California Association of Governments (SCAG). 1987. The Path to Clean Air: Attainment Strategies. December. Los Angeles, Calif.: SCAG.
South Coast Air Quality Management District (SCAQMD) and Southern California Association of Governments (SCAG). 1988. The Path to Clean Air: Policy Proposals for the 1988 Air Quality Management Plan Revision. June. Los Angeles, Calif.: SCAG.
Shiller, J. W. 1987. The role of alternative fuels in air quality planning: Methanol fuels in light duty vehicles, does it help or hurt? Proceedings, Air Pollution Control Association Specialty Workshop on Post-1987 Ozone Issues, Golden West Chapter, San Francisco, November.

Staner, H. W., ed. 1905. Alcohol as a fuel for motor cars. The Autocar XIV(482)(January 14).
Szwarc, A., and G. M. Branco. 1987. Automotive emissions—The Brazilian control program. SAE Technical Paper Series, SAE-871073. Warrendale, Pa.: Society of Automotive Engineers.
U.S. Department of Energy. 1988. Assessment of Costs and Benefits of Flexible and Alternative Fuel Use in the U.S. Transportation Sector, Progress Report One: Context and Analytical Framework, DOE/PE-0080. Washington, D.C.: U.S. Department of Energy.
U.S. Energy Information Administration. 1988. Monthly Energy Review (July). Washington, D.C.: U.S. Department of Energy.
U.S. Environmental Protection Agency (EPA). 1977. Uses, Limitations and Technical Basis of Procedures for Quantifying Relationships Between Photochemical Oxidants and Precursors. EPA-450/2-77-021a. EPA Research Triangle Park, N.C.
U.S. Environmental Protection Agency. 1985. Outdoor Smog Chamber Experiments: Reactivity of Methanol Exhaust. EPA Office of Mobile Sources. EPA 460/3-85-009a & b. September.
U.S. Environmental Protection Agency. 1986a. Regulatory Support Document, Proposed Organic Emission Standards and Test Procedures for 1988 and Later Methanol Vehicles and Engines. July.
U.S. Environmental Protection Agency. 1986b. Emission Standards for Methanol-Fueled Motor Vehicles and Motor Vehicle Engines, FR 51 No. 166. August.
U.S. Environmental Protection Agency. 1987a. Air Quality Benefits of Alternative Fuels. Report prepared for the Vice President's Task Force on Alternative Fuels, EPA Office of Mobile Sources. July.
U.S. Environmental Protection Agency. 1987b. Guidance on Estimating Motor Vehicle Emission Reductions from the Use of Alternate Fuels and Fuel Blends. Draft Technical Report EPA-AA-TSS-PA-87-4. EPA Emission Control Technology Division, Ann Arbor, Mich. July.
U.S. Environmental Protection Agency. 1987c. Note to correspondents, EPA release of 1986 air quality data. EPA Office of Air Quality Planning and Standards. August 27.
U.S. Environmental Protection Agency. 1988a. National Air Quality and Emission Trends Report, 1988. EPA-450/4-88-001.
U.S. Environmental Protection Agency. 1988b. EPA lists areas failing to meet ozone or carbon monoxide standards. EPA press release, Washington, D.C., May 3.
U.S. Environmental Protection Agency. 1988c. Guidance on Estimating Motor Vehicle Emission Reductions from the Use of Alternative Fuels and Fuel Blends. EPA-AA-TSS-PA-87-4. Ann Arbor, Mich.: EPA Emission Control Technology Division.
Whitten, G. Z., and H. Hogo. 1983. Impact of methanol on smog: A preliminary estimate. Systems Applications, Inc., San Rafael, Calif., for ARCO Petroleum Products Co., SAI Publication No. 83044. February.
Whitten, G. Z., T. C. Myers, and N. Yonkow. 1986. Photochemical modeling of methanol-use scenarios in Philadelphia. Systems Applications, Inc., San Rafael, Calif., for EPA Emission Control Technology Division, Ann Arbor, Mich. EPA 460/3-86-001. March.

3
Evolving Vulnerabilities and Opportunities

Managing Volatility in the Oil Industry

JOHN F. BOOKOUT

A concern often expressed nowadays is that the survivability of parts of the oil industry is threatened. Over the past 15 years the oil industry has experienced a succession of rapid and significant changes. The dimensions of change include oil price, of course, but also changes in activity levels, employment, supply and demand, and in economic and political conditions. In this chapter we will examine some of these changes.

First, we will compare the volatility of oil prices with that of other commodities, to determine whether the oil industry environment is unique in any way. Also in this discussion we will consider the impact of futures market trading. Second, we will look at the performance of various parts of the industry—the upstream exploration and production sector, the refining and retailing parts of the downstream sector, as well as a brief look at consumers. Last, we will consider whether the industry has adapted to this changed environment, and what we might expect in the 1990s.

LONG-TERM COMMODITIES PRICES

Although many in the oil industry may not think of it this way, the experience over the past 15 years, in fact over the past 60 years, has many precedents. A statistical analysis of the price history for corn, soybeans, tin, and oil indicates that price volatility of oil is similar to that of other commodities (see Figure 1). The same picture emerges if oil prices are superimposed on plots of copper, sugar, cocoa, rubber, and many other commodities. It is also useful to note that there have been several epochs in

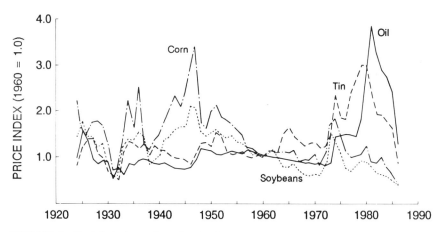

FIGURE 1 Real (inflation adjusted) price indexes for corn, soybeans, tin, and oil (1960 = 1.0).

these various price series. In particular, the period from the early 1950s to about 1970 tended to be rather calm for many commodities, including oil. Before 1950 and after 1970, prices tended to be unstable. Statistical testing of the oil price series, including comparison with the other commodities, also indicates two other points: until 1973, price volatility in oil was less than in other commodities, but only marginally so; after 1973 volatility in oil prices was greater than in all other commodities.

These facts suggest that the 20-year period of relative tranquillity before 1973 was just as abnormal as the 15-year period of instability after that date. We now shall argue that the oil industry has been adapting itself in a way that is likely to accommodate, and could moderate, volatility in the future.

But before leaving the subject of commodities prices, let us address an issue dealing with futures markets. Some people have suggested that the introduction of trading in oil futures has been a major cause of the volatility evident in the market today.

FUTURES TRADING AND PRICE VOLATILITY

Most people would agree that day-to-day oil price movements are now more volatile than they were 20 years ago. At that time, most transactions were based on contract or other long-term relations. But in the context of prospects for the 1990s, have year-to-year price changes been affected?

A study done recently by Shell analyzed the price volatility of 22 different commodities, before and after the start-up of futures trading (Bookout, 1988). From the results, we concluded that for most commodities, including

TABLE 1 Effect of Futures Trading on Price Volatility

Commodity	Initial Trading Date	Effect
Metals		None
Aluminum	1965	
Copper	1974	
Lead	1965	
Nickel	1969	
Tin	1965	
Zinc	1965	
Grains		None
Maize	1859	
Rice	1981	
Sorghum	1965	
Wheat	1859	
Other edibles		
Increases		
Beef	1965	
Cocoa	1925	
Coffee	1955	
Sugar	1941	
Soybeans	1936	
Meal	1951	
Oil	1970	
Agricultural raw materials		None
Cotton	1981	
Palm oil	1975	
Rubber	1965	
Wool	1965	
Crude oil	1983	None

oil, there was no more volatility after futures trading began than before, as measured by the variance in quarterly prices. Oil futures markets, like other futures markets, appear to be a response to, rather than a cause of, price volatility (see Table 1).

As we will see later, the establishment of oil futures is just one example of how the oil industry has embraced the technology of other industries, in this case, financial markets, as a way of adapting to an environment of volatile prices.

U.S. OIL AND GAS ADDITIONS AND PRODUCTION

Figure 2 shows the history of U.S. oil and gas production and additions to the known reserve base in equivalent barrels per year. Additions are

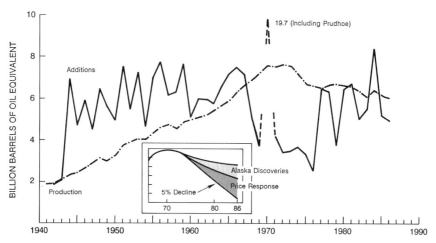

FIGURE 2 U.S. oil and gas production and additions to reserves (billion barrels of oil equivalent). The inset shows the reserve additions due to discoveries in Alaska and response to price increases.

defined as the volumes of hydrocarbon added to the inventory of proven reserves through discoveries, development drilling, and revisions of the volumes of existing recoverable oil and gas deposits as more is learned about reservoir properties over time.

As the figure shows, during the period from 1940 to the late 1960s there was a large backlog of opportunities to add to the country's reserve base, even though there were low prices prevailing at the time. Additionally, the regulatory bodies of the various states had proration policies that limited the oil production rates below capacity. For these reasons, until the late 1960s, additions were well above production. By the end of the 1960s, the backlog of low-cost prospects had been depleted, additions were down, and so reserve replacement was in decline.

In 1970 the United States was fortunate to have added to its reserve base the giant field at Prudhoe Bay, Alaska. But reserve replacement in the lower 48 states continued to fall off, and production started to decline by 5 percent after 1970. It was not until the stimulation of the price increases in 1973 that reserve replacement picked up. The inset in Figure 2 shows that, if the 5 percent decline that was apparent after 1970 continued indefinitely, oil and gas production today would be nearly 40 percent less than it is, that is, about 6 million barrels of oil equivalent per day less.

The data suggest that increase in additions to reserves, stimulated by the price increases in the 1970s, contributed about 65 percent of what was needed to stabilize production. Alaska contributed the remainder, and many would suggest that Alaskan production could be brought to market only because of the price increases. In both cases, the maximum volume effects came six to eight years after the price change.

These statistics probably do not give a sense of the role that technology played over this period. The advances in the sciences of geology and geophysics before 1970 are now taken for granted. But where would the industry be without the breakthroughs in such fields as seismology, stratigraphy, and plate tectonics? Those tools added to our routine vocabulary such terms as *bright spots, source rocks,* and *lithofacies,* to name a few. In more recent years, particularly in response to the imperatives perceived in the 1970s, advances in three-dimensional seismic imaging, enhanced recovery techniques, and dealing with harsh environmental conditions such as deep water in the Gulf of Mexico and ice packs in the Beaufort Sea, have made available oil and gas resources that were sometimes not even considered feasible 20 years ago.

EXPLORATION AND PRODUCTION SECTOR ACTIVITY

Figure 3 presents several key indicators of oil industry activity. Three phases of wellhead price behavior can be identified:

- 1960–1973: oil and gas prices were constant or declining
- 1973–1981: prices increased rapidly
- 1981–1986: prices declined

Three key activities over this period seem to follow closely the oil and gas price profiles—the number of seismic crews employed in the United States, the number of rotary rigs in operation, and the number of wells completed. Each of these activities responds to the price pattern of gradual decline, surge, and collapse. There have also been changes in efficiency and productivity in response to this cycle.

Behavior Patterns

The number of well completions per active drilling rig is a measure of efficiency in drilling. During the era of declining prices in the 1960s, the level was about 24 to 28, as shown in the left graph of Figure 4. After price increases in the 1970s, the number of wells drilled increased, but the number of completions per rig declined about 20 percent. Reasons for this decline include lack of experience among the new drilling crews, lack of good matching between rigs and their prospects, and the fact that some older rigs may have been kept in service longer than they should have been. More recently, oil prices have come down and the utilization of the fleet of rigs has declined; productivity of the operating rigs has

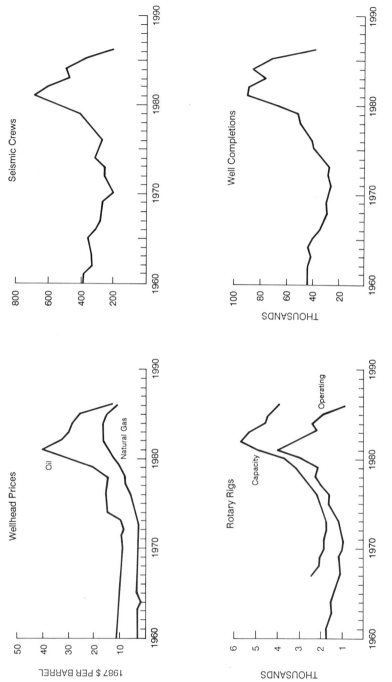

FIGURE 3 Four key indicators of U.S. oil industry activity: wellhead price, number of seismic crews, rotary rig capacity and utilization, and number of wells completed.

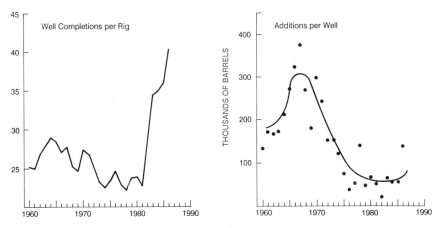

FIGURE 4 Trends in exploration performance by U.S. oil industry.

soared, as the industry responded to the need to improve productivity during retrenchment.

The graph on the right of Figure 4 shows another measure of behavior, the size of the hydrocarbon additions per successful well drilled. In the 1960s industry was upgrading by selecting higher volume prospects as prices gradually declined. With the price rises in the 1970s, the additions per well dropped as the incentives to drill smaller prospects increased. In 1986, the last year for which we have data, the size of the additions per well rose, suggesting that industry has once again started to upgrade its prospects.

Revenues

Turning now to the financial behavior of the industry, Figure 5 shows the sources of industry revenues, which peaked in 1981 at $180 billion. On the right, the percentage disposition of those revenues is divided among the royalties and production taxes and the costs of lifting, manpower, and materials (overhead). The remainder is operating cash income, which is the cash left for reinvestment, paying income taxes, dividends, or debt. The wedge labeled WPT represents the revenues consumed by the windfall profits tax.

Over the 26-year history shown in Figure 5, operating cash income has been a remarkably consistent 60 percent of total revenues, with the exception of the period when some $80 billion was paid in windfall profits tax.

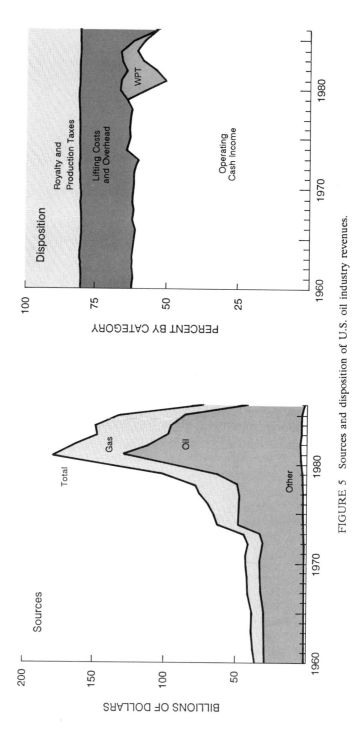

FIGURE 5 Sources and disposition of U.S. oil industry revenues.

Reinvestment Patterns

Reinvestment includes both those costs that can be expensed, such as dry hole costs, some drilling costs, research, and so forth, as well as the capital expenditures for exploration and production. In the left graph of Figure 6, the total funds reinvested in exploration and production are compared with the operating cash income. The right graph shows the ratio of reinvested funds to operating cash income. Despite the unstable market that prevailed, and the intrusions of the federal government, the expansion and contraction behavior of the industry results in a rather stable ratio of about 70 percent.[1]

Incidentally, the low point in 1971 was due to the absence of any federal lease sales in that year. And the high point in the early 1980s reflects the high price expectations that prevailed. Coupled with the behavior shown in these last few figures, this consistent pattern of reinvestment throughout the period is a sign of continuing adaptation to the changing environment.

FUTURE CRUDE OIL PRICES

Turning to the future, let us look at the kind of volatility that might be expected. In a recent report by the National Petroleum Council (NPC, 1987), two oil price trajectories were used as guidelines for their study. In this chapter, the NPC price projections, shown in Figure 7, are used as boundary conditions for the variation in year-to-year prices. Crude oil prices in the year 2000 are projected to range from $20 to $35 per barrel, in 1987 dollars. Consistent natural gas prices are also assumed by equilibrating gas with fuel oil prices sometime in the early 1990s. To model the future performance of the U.S. oil industry, each boundary will be used as a price trajectory in conjunction with the industry behavior just described in the previous figures: expansion and contraction capability, change in efficiencies, and financial performance.

FUTURE U.S. OIL AND GAS ADDITIONS AND PRODUCTION

The logic of the model that underlies the projections on Figure 8 is that the price in any year determines revenues, which in turn control investment level. After accounting for price effects on efficiency, productivity, finding rates per well, and additions to reserves, production for the following year is determined.

With the model structured as described, depending on the price path, additions to reserves would fall to the level of the early 1970s, that is, the 3- to 5-billion-barrel range, and the rate of production would approach the rate of additions. With no surprises, U.S. oil and gas production would decline 1 to 3 percent per year.

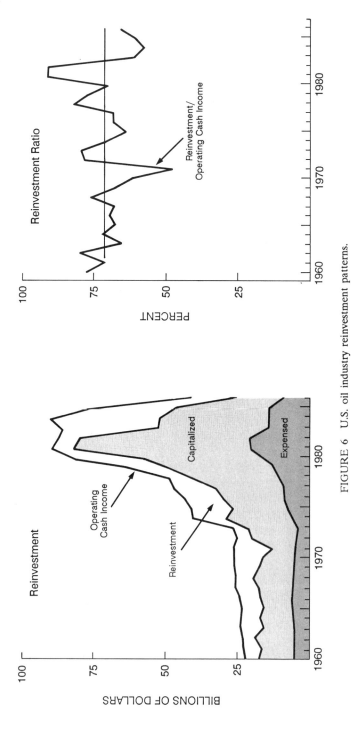

FIGURE 6 U.S. oil industry reinvestment patterns.

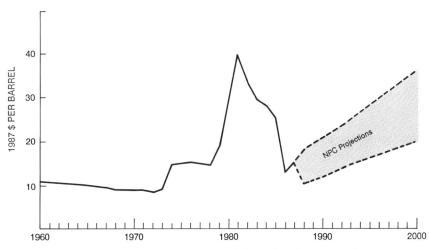

FIGURE 7 U.S. crude oil price history and National Petroleum Council projections to the year 2000.

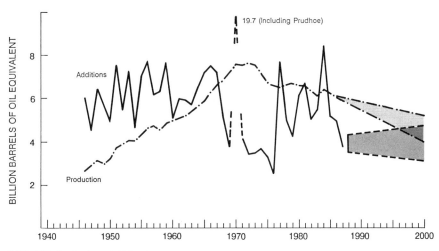

FIGURE 8 Projected U.S. oil and gas production and additions to reserves (billion barrels of oil equivalent).

TABLE 2 U.S. Discovered Oil

Oil Resources	Total Oil (billions of barrels)	Enhanced Recovery Oil	
		Billions of Barrels	Percent of Resource
Produced	146	3	2
Proven	30	4	13
Probable	21	8	38
Total	197	15	
Oil left in place	362		

What this model does not reflect are the statistically less probable events such as Prudhoe Bay, which in the 1990s could substantially alter the oil and gas production profile.

GEOLOGIC AND TECHNOLOGICAL SURPRISES

The kind of change not reflected in the model could come as a geologic surprise capable of sustaining industry production at present or even higher levels. However, some initial period of decline is likely before any "surprise" can be brought into production. A surprise could be rooted in technological advances or even breakthroughs, which could release large resources of oil and gas that are not now considered economic. Tight sands gas, coal seam gas, shale oil, and enhanced oil recovery techniques are possibilities.

To illustrate the advances in enhanced oil recovery, consider that a total of 197 billion barrels is expected to be produced from oil resources already identified, as shown in the left column of Table 2. That represents an overall recovery rate of about 35 percent of the oil in place, leaving 362 billion barrels behind. Present enhanced recovery technology has already boosted our yield by about 15 billion barrels, as shown in the center column of the table. If that overall recovery rate could be increased from 35 percent to 45 percent, we would recover another 36 billion barrels from the oil expected to be left in place, in effect doubling our proven reserves.

What is encouraging in this regard is the behavior that industry has exhibited in its commitment to research. An average of more than half a billion 1987 dollars has been committed to research annually since the early 1970s. Technological milestones continue to be achieved in areas such as enhanced oil recovery, deep-water exploration and development, and management of harsh environments.

EXPLORATION AND PRODUCTION SECTOR REVENUES AND CASH FLOW

The cash position of the oil industry has changed over the years. On the top graph of Figure 9 are past and projected revenues from 1960 to 2000. On the bottom graph are the net cash flows associated with those revenues and the projected financial performance based on the model.

During the 1960s the industry was generating surplus cash, even after paying out dividends. By the 1970s a combination of higher tax rates and higher capital spending rates drove net cash flow nearly to zero. In fact, the industry took in outside funds to support its activities. In the future, under the model assumptions, the revenues vary, both because of the different price trajectories and because of the different oil and gas production responses. By the year 2000, industry revenues could nearly return to the $180 billion level we have already seen.

Perhaps more important to the viability of the industry under changing conditions is the projection of net cash flow. Even at the lower boundary of the NPC projections shown in Figure 7, the industry maintains a positive cash flow, in part by lowering the absolute level of reinvestment in the lower

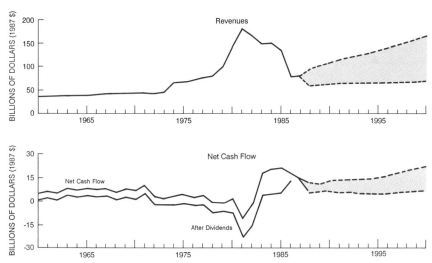

FIGURE 9 Historical and projected financial performance of the U.S. oil industry (billions of 1987 dollars).

price environment. At both boundaries it would appear that dividends can continue to flow at historical levels.

This model leads to a number of observations. If the behavior patterns that the industry has exhibited in the past continue in the future, the industry will continue to be financially viable. A moderate decline of the U.S. oil and gas production is likely based on our analysis of the NPC model. However, there is reason to be hopeful about eventual improvement in that position on the basis of some geologic and technological precedents. In any event, it appears that the upstream sector is resilient in its ability to accommodate large, rapid changes in its business environment.

Other forms of volatility in the oil industry are important measures of the ability to adapt in sectors such as refining.

U.S. CRUDE OIL IMPORTS

The month-to-month variation in imports of crude oil from 20 different countries supplying almost all the U.S. imports is shown in Figure 10. By inspection, there seems to be more chaos at the middle and end of the period than at the beginning. Typical month-to-month changes in the exporting countries' receipts nearly tripled from 1971 to 1979, from plus or minus 1 million barrels to plus or minus 3 million barrels; since then, volatility has changed very little (U.S. Department of Energy, 1988).

The point of focusing on this measure of volatility is to consider the response of U.S. refiners to this changing environment. The quality of crude oil varies from source to source, requiring different refining facilities to optimize conversion of the raw material to finished oil products. As the

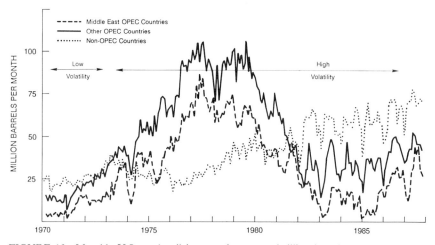

FIGURE 10 Monthly U.S. crude oil imports, by source (million barrels per month).

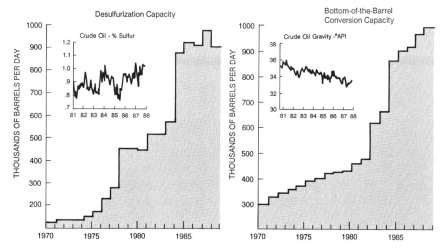

FIGURE 11 Capacity of U.S. Gulf Coast refineries (thousands of barrels per day). The inset in the left graph shows a slowly rising average sulfur content of crude oil, which is matched by rising desulfurization capacity of refineries. The inset in the right graph shows decreasing average refinability as measured by API gravity. This decrease is matched by rising bottom-of-the-barrel (heavy crude oil) conversion capacity.

availability of crude from any source has become more volatile, refiners seem to have positioned themselves to handle a broader range of quality.

CHANGES IN REFINING CAPACITY

Much of the imported crude oil in the United States is processed in refineries on the Gulf Coast. The left graph of Figure 11 shows the increase in capacity to remove sulfur from crude oil. Nearly a million barrels per day of desulfurization capacity has been added in these refineries since 1970. Detailed data on the actual sulfur content of the crude oil being processed in these refineries are available only as far back as 1980. But just since that time, the sulfur content of the crude oil slate processed on the Gulf Coast has, on average, increased from 0.8 to 1.0 percent, as shown in the inset. This increase has occurred despite the fact that there are more lower-sulfur crude oils being produced in the world today than in 1980. Although refiners are now less likely to know what crude oil will be available, they appear to have increased their ability to be competitive in the light of this uncertainty.

As shown on the right graph of Figure 11, facilities to process the bottom-of-the-barrel crude oil have been increased. The API gravity of a crude oil is one measure of its quality, since it indicates the amount of very heavy hydrocarbons contained in the crude.[2] Without sophisticated refining equipment, a refinery will produce only lower-valued heavy fuel oil from a lower-gravity crude oil, if it can process it at all. In the period shown, more than 600,000 barrels per day of capacity has been added to the Gulf Coast refineries to convert this bottom-of-the-barrel crude oil into light products, primarily transportation fuels. Nationwide, more than $4 billion has been spent in the past six years for these types of upgrading facilities.

As the inset on the right graph of Figure 11 shows, this upgrading has permitted refiners to select crude oils that are increasingly heavy, as measured by their declining gravity. This change is in response to a generally increased, but more unstable, availability of lighter crude oils in the world.

In sum, these two graphs indicate ways in which another sector of the oil industry has adapted itself to a changed business environment of increased volatility. They have sensed the need for changing the way they do business, and have developed or absorbed the technology necessary to accomplish it.

RETAIL GASOLINE SECTOR

Downstream of the refining sector, the retail gasoline business has adapted to change in a somewhat different way. The mid-1970s might be considered a watershed in the perceptions of consumer needs in gasoline retailing. As consumers saw gasoline prices increase, their requirements shifted from full service to economy. They were, in fact, deflecting the effects of the unstable prices back to the retail sector. As a consequence, although less than 25 percent of the gasoline sold was self-serve in 1976, nearly 80 percent of industry sales are self-serve today, as seen on the left graph of Figure 12. The right graph shows that after 1975, the number of service stations in the United States declined 4 percent annually, from 190,000 to 117,000 in 1987. But during that period of declining total station count, the throughput of the average station increased from 32,000 gallons per month to 55,000 per month (Lundberg Letter, July 17, 1987). The retail gasoline sector has emerged from this recent period of volatility with substantially more productive facilities than it had 15 years ago. The change has come about through a combination of innovative management techniques, technological advances in electronics, and improved design.

FUEL SWITCHABILITY IN INDUSTRIAL BOILERS

One final example of the changes brought about by price volatility is not in the oil industry itself, but does have an impact on the industry. Energy consumers have done much to accommodate their vulnerability to instability in both price and volume. In the case of industrial energy consumers, the

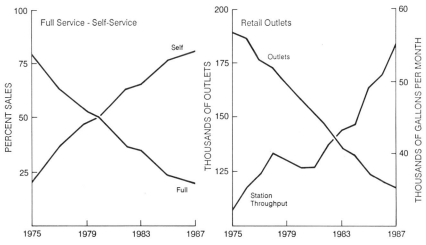
FIGURE 12 Trends in U.S. gasoline retailing.

standard policy seems to be to install dual fueling capability in any major new installation. Many existing facilities have even been retrofitted in this manner. Figure 13 shows the switchability by fuel and the quantity of fuel that can be switched by several major industries. Industrial consumers, much like refiners, have adapted their capital facilities to accommodate changes in both price and volume. These consumers are no longer captive customers.

This capability of consumers to switch away from oil, and refiners to switch among sources of crude, suggests that supply disruptions like those experienced in both oil and gas in the 1970s might not have as severe an impact the next time around. Consumers now have more flexibility.

A SYSTEMS THEORY APPROACH

To draw conclusions from these analyses, some concepts from general systems theory might be borrowed. In that discipline, the oil industry might be thought of as an "open system" (Figure 14) that has to interact with its environment to remain viable. For the purposes of this discussion, the U.S. oil industry can be regarded as an open system that includes the companies (upstream exploration and production sectors and downstream refining and marketing), the consumers, and the related service industries (e.g., oil field, financial). That system can be contrasted with many mechanical feedback systems, from a simple thermostat to a sophisticated expert system. These are closed systems and do not have the ability to rejuvenate themselves.

When an environment changes abruptly, the initial reaction of an

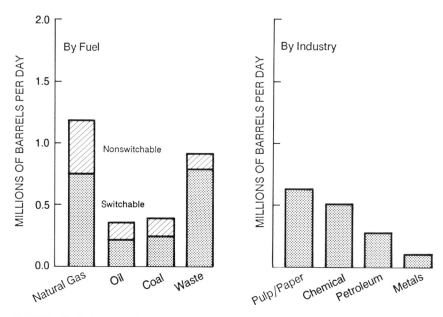

FIGURE 13 Industrial boiler fuel switchability among industrial users (million barrels per day). SOURCE: Petroleum Industry Research Associates (1986).

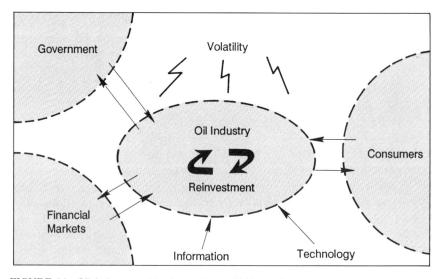

FIGURE 14 Oil industry and its interactions with its environment.

open system is often survival, by shutting out these changes. For long-term viability, the system must open up again and adapt. In the oil industry, the environment has changed with the increase in volatility.

At the same time, the response capability of the industry was curtailed as the government encroached on the industry's boundaries. We saw the impact of windfall profits tax on the ability of the upstream segments of the industry to invest. Yet we have seen a number of examples of how the industry began to adapt. As the industry continued to develop its own technology internally, it adopted innovations from other disciplines— computer science, engineering, and so on. The industry employed financial technology in the form of futures trading as it sensed the need to externalize some of the financial risks. The oil industry even changed the shape of its own environment, as industrial and retail consumers positioned themselves to accommodate a volatile oil market.

From this analogy, and from the other observations already made, it seems clear that to remain viable the oil industry must continue to position itself to handle whatever the business environment presents. Some might feel that the industry ought to insulate itself from volatility, but that does not make the volatility go away. In the long run it only deflects the volatility somewhere else. If that target, which might be consumers, adapts to the environmental change, it becomes more viable, and the oil industry probably loses ground. Consider, for example, how much less oil and gas consumers need now than they did 15 years ago. U.S. consumption of oil and gas per dollar of real GNP has dropped to 62 percent of its 1973 level.

CONCLUSIONS

From this discussion, together with the studies that have been presented here, it appears that the oil industry has demonstrated considerable ability to adapt to a world of uncertainty and volatility. The refining sector seems to have adopted the position that the best approach to volatility is flexibility. The upstream sectors of the industry have exhibited substantial ability to expand and contract.

From a financial point of view, the oil industry can be expected to be a survivor and to be robust over a broad range of futures. And finally, although we might expect a gradual decline in U.S. oil and gas production in the future, there are reasons to hope that new and ingenious applications of technology in the coming years will favorably improve the outlook.

It seems that there is hope for somewhat more moderate fluctuations in the oil industry's future than it has experienced in the recent past. But to remain viable for the long term, the oil industry must continue to sense the need to reshape itself through open, continuing interaction with its environment.

NOTES

1. The real (inflation adjusted) rate of return has varied over the period from highs in 1973 and 1977 of 8 and 9 percent to lows slightly below zero in 1982–1983. During the 1960s, real rates of return averaged about 4 percent; in the 1970s, about 6 percent. The drop in oil prices has reduced the real rate of return of investments made in the early 1980s to zero or below, given current price expectations.
2. API gravity is the oil industry measuring standard and is inversely proportional to specific gravity in the following way:

$$°API = (141.5/\text{specific gravity}) - 131.5$$

REFERENCES

Bookout, J. F. 1988. Impact of Volatility on the U.S. Oil Industry. Background studies. Houston, Tex.: Shell Oil Company.

Lundberg Letter. July 17, 1987. Vital Statistics and Analysis in Oil Marketing and Related Industries. North Hollywood, Calif.: Lundberg Survey, Inc.

National Petroleum Council. 1987. Factors Affecting U.S. Oil and Gas Outlook. Washington, D.C.: National Petroleum Council.

Petroleum Industry Research Associates. 1986. The U.S. Industrial Energy Market Outlook: Demand Prospects, Interfuel Competition, Strategic Implications, Vol. 1. New York: Petroleum Industry Research Associates, Inc.

U.S. Department of Energy. 1988. Monthly Energy Review. Washington, D.C.: U.S. Department of Energy.

The Uncertain Future Role of Natural Gas

WILLIAM T. McCORMICK, JR.

Although the future role of natural gas in the U.S. energy picture is not certain, it is probably no less certain, and quite possibly more certain, than other energy fuels available in the United States in the next 20 to 30 years. We know this because natural gas ranks high among energy fuels when it is evaluated against the key criteria that will determine its competitiveness in the national energy fuels marketplace.

Those key criteria are the size of the recoverable resource, the economics of both extraction and use, and public acceptance with respect to environmental, safety, and siting matters. This chapter discusses each of these points separately and then summarizes the future contribution of natural gas in the U.S. energy mix.

The future contribution of natural gas will depend on the viability of other fuels in the marketplace. And since each fuel type has its own challenges, the equation certainly has more than one variable.

As a utility executive and a nuclear engineer, I strongly support nuclear and coal development, but I also understand the substantial and real challenges to the development of these resources: nuclear in the short term and coal, because of concerns about sulfur dioxide and carbon dioxide, in the longer term. To the extent that one or more energy resources are not developed in the normal economic pattern, this necessarily implies that some competitive resources must bridge the gap. Unfortunately, in the United States, because of political and other factors, the question is not necessarily which energy resource is best, but rather which might be least objectionable. On to natural gas.

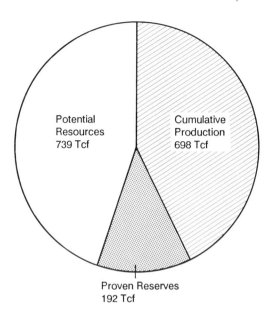

FIGURE 1 U.S. natural gas cumulative production, reserves, and potential in trillions of cubic feet (Tcf) as of December 31, 1986.

RESOURCE AVAILABILITY

Figure 1 shows potential resources of natural gas are approximately four times larger than current proven reserves. Potential resources are gas volumes that are estimated to be recoverable, although not currently economical.

Although natural gas has been in short supply in the past, it appears that, as a result of wellhead price deregulation, past shortages were due almost entirely to low regulated wellhead prices, which created a noneconomic exploration and development environment (Gibson, 1988). Figure 2 shows that additions to natural gas reserves dropped when the level of development activities, as indicated by the annual number of well completions, responded to the disincentives. In fact, since 1977, when wellhead gas prices for almost all newly discovered natural gas were first deregulated, the additions to new gas reserves in the United States have averaged about 90 percent of annual production. At this replacement rate, U.S. gas reserves would satisfy current levels of consumption for about 50 years.

Moreover, the Potential Gas Committee, an industry, government, and academic group, estimates that in addition to the 160 trillion cubic feet (Tcf) of proven U.S. gas reserves, there are an additional approximately 650 Tcf of potential U.S. gas resources (see Table 1); that is, gas that is in place but not recoverable at current prices. Together these gas resources and reserves amount to about 50 years of supplies at the current annual U.S. consumption rate of 18 Tcf.

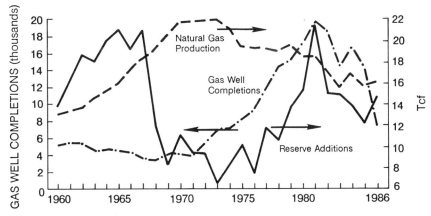

FIGURE 2 Natural gas well completions, reserve additions, and production in the conterminous United States.

TABLE 1 Remaining Lower-48 Gas Resources as of January 1, 1987 (Gas Research Institute Baseline Projection)

Category	Resources (Tcf)
Proven reserves[a]	158.9
Reserve appreciation	149.5
Undiscovered new fields	265.0
New technology increment[b]	227.6
Total	801.0

[a] U.S. Department of Energy (excludes Alaska).
[b] Incremental resource for tight sands and Devonian shale only.

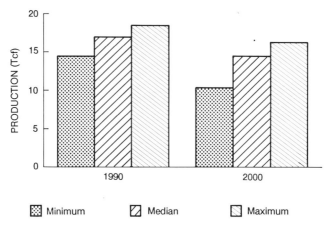

FIGURE 3 Conventional production of natural gas in the conterminous United States (20 estimates).

These estimates of gas resources do not fully account for the effect of expanding the recoverable resource base from the use of new or improved recovery technologies. Just as new technologies have added significantly to domestic petroleum reserves (see Bookout, in this volume), new technologies could add hundreds of Tcf of new reserves, resulting in several decades of additional U.S. domestic gas resources.

None of the preceding discussion assumes any contribution from Canadian gas, which is largely untapped, readily available, and secure. In fact, Canadian reserves and resources of natural gas amount to about 500 Tcf and would almost double the resource base available to the United States. Given the new free-market approach of the Canadian government toward gas exports, and the likelihood of ratification of the Canadian/United States Free Trade Agreement, it is increasingly clear that Canadian natural gas resources will be an important supplement to U.S. gas supplies.

In short, from a physical resource standpoint, natural gas is an energy resource that should be available for many years to come. Estimates of future gas production based on a consensus of government, industry, and academic studies range between about 15 Tcf and 18 Tcf in 1990 and between 10 Tcf and 16 Tcf in 2000 (see Figure 3). As shown in Table 2, the total available U.S. supply of natural gas exceeds current use for the present and immediate future.

ECONOMICS OF NATURAL GAS

The key question relating to the future contribution of natural gas in the next 20 to 30 years is not one of available resource base, but one of economics; namely, is the supply curve for natural gas such that it can compete in the various end-use sectors?

The answer to this question, of course, depends as much on the costs of, and technological developments that influence, resource extraction as on transportation economics and end-use efficiencies. Figure 4 shows the results of a Gas Research Institute-funded study that concludes that with existing technology about 400 Tcf of gas resource can be recovered at $6.00 per million Btu (MBtu) or less. With improved technology, the recoverable resource can be increased to 700 Tcf at the same price.

From the standpoint of end-use efficiencies, there are several recent applications and developments that can make natural gas an increasingly economical fuel. These developments include natural gas vehicles, gas cofiring in boilers with coal, and combined-cycle gas power generation.

Table 3 shows the current relative economics between natural gas and gasoline for vehicle use. It shows that *today* natural gas delivered retail equates to only about 60 cents per gallon of gasoline. Thus, there now appears to be sufficient operating savings to offset the capital cost of vehicle conversion.

TABLE 2 U.S. Gas Supplies, Gas Research Institute Baseline Projection (quads), 1987

Production Basis	1986	1990	2000	2010
Current practice				
Domestic production	16.6	16.3	14.3	9.3
Canadian imports	0.8	0.9	1.4	1.1
Liquefied natural gas imports	a	0.1	0.3	0.8
Supplemental sources	a	0.2	0.3	0.3
Total	17.4	17.5	16.3	11.5
New initiatives				
Lower-48, advanced technology	0.0	0.3	2.6	5.4
Alaskan pipeline	0.0	0.0	0.0	1.2
Canadian frontier	0.0	0.0	0.5	0.7
Other imports	0.0	0.0	0.0	1.0
Synthetics	0.0	0.0	0.0	0.1
Total	0.0	0.3	3.1	8.4
Total supply	17.4	17.8	19.4	19.9

NOTE: 1 quad = 0.979 Tcf.

a Less than 0.05 quad.

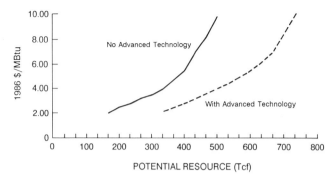

FIGURE 4 Effects of advanced technology on cost of potential gas resources in tight sands, Devonian shale, in the conterminous United States (as of January 1, 1981).

Cofiring involves the use of small amounts of natural gas (generally 5 to 15 percent of the boiler heat input) with coal in utility boilers. Because natural gas burns cleanly, cofiring of relatively small amounts of gas with coal in utility boilers will (1) directly reduce emissions of nitrogen oxides (NO_x); (2) increase the productivity and operating flexibilities of boilers; and (3) extend significantly the reserves of environmentally acceptable coals that can be burned by the utilities while meeting proposed acid rain legislation at one-third the cost of using scrubbers.

Another important advantage of gas cofiring is to complement coal in conventional and advanced clean coal technologies for pollution control.

TABLE 3 Relative Economics of Natural Gas versus Gasoline as Vehicle Fuel, 1988

Fuel	Cost ($)
Natural gas	
$/MBtu	4.50
$/Barrel	26.00
$/Gallon	0.60
Gasoline	
$/Gallon	1.00

Gas has negligible sulfur and no particulate content and can be used as a reburn fuel to reduce NO_x emissions by 50 to 60 percent. Consequently, gas can be used in small amounts with coal to help control pollution in marginal compliance situations, such as when the sulfur content of the coal is out of specifications (sulfur trim) or the electrostatic precipitator (to remove particulates) is not operating properly. Gas can also be cofired with higher sulfur coals to meet sulfur dioxide compliance limits being considered in acid rain legislation and to extend low-sulfur coal reserves.

A most important aspect of the above elements of gas cofiring is that gas does not displace coal. The use of gas would enhance coal combustion by improving efficiency or by solving a problem with slagging, opacity, or system reliability or by helping the boiler achieve peak load.

A fine example is the use of natural gas for electricity generation, which 5 to 10 years ago was considered unthinkable. However, today, many utilities are not only considering it but believe it to be a first-choice option.

How can this be when the cost of delivered coal is between $1.50 and $2.00 per MBtu and delivered gas is $3.00 to $3.50 per MBtu? The answer is that new gas turbine technology used in the combined-cycle mode with improved steam turbines has allowed 90 percent plant availability and 43 percent plant efficiency for combined-cycle gas plants compared with 80 percent availability and 36 percent efficiency for modern coal plants. This in conjunction with a 40 percent lower capital cost for a gas-fueled electric plant versus a coal plant (in part because of the absence of certain environmental costs) results in a situation in which a utility can pay 50 to 100 percent more for natural gas fuel than for coal on a $/MBtu basis to get the same end-user costs for electricity generation (see Table 4). In addition, because of the short construction time and lower capital cost of gas plants, utilities perceive less financial risk with a gas-fueled plant.

ENVIRONMENT, SAFETY, AND SITING

One of the most serious challenges facing all energy projects today relates to public acceptance. From this standpoint, natural gas ranks high. As a clean-burning fuel with minimal pollution, gas is the cleanest of the fossil fuels. Although, in contrast to nuclear power, it does have some modest air pollutant emissions, it does not share some of the difficult public acceptance challenges of nuclear power, namely, in plant safety and waste disposal. Moreover, unlike coal, gas does not suffer some of the problems that require the use of scrubbers and other expensive equipment in power plants.

Another consequence of public acceptance problems for nuclear and, to some extent, coal plants is the long delay in authorization and construction of new plant facilities. This is becoming increasingly important in

TABLE 4 Comparison of Costs of Electric Generation Fueled by Coal and Gas

Cost Factor	Coal	Gas (combined cycle)
Capital cost ($/kW)	$1,800	$1,000
Thermal efficiency (%)		
Existing technology	36	43
New technology	37	46
Plant availability (%)	80	90
Construction time (years)	6 – 8	3 – 4
Current fuel cost delivered ($/MBtu)	$2	$3

utility decision making because of the increasing prevalence of State Public Utility Commission prudency investigations (see Balzhiser, in this volume) of plants that experience construction delays and cost overruns. Because gas plants can be built faster and less expensively than either nuclear or coal plants and are relatively clean, this represents an advantage for natural gas.

CONCLUSION

In summary, although the future for natural gas in the national energy mix is by no means certain, it does offer some attractive advantages. The economic savings resulting from these advantages appear to be substantial enough to offset the increasing price of natural gas for some time in the future. How long into the future this will last depends to a large extent on new, emerging technologies not only for natural gas but also for its competing fuels in the marketplace.

REFERENCE

Gibson, D. E. 1988. The United States natural gas industry: The transition to a free market. Paper presented at 17th World Gas Conference, Washington, D.C., June 5–9 1988. Order no. IGU/J1-88. Washington, D.C.: International Gas Union.

European Natural Gas Supplies and Markets

HENRIK AGER-HANSSEN

During the past 15 years a remarkable shift in the composition of the world hydrocarbon reserves has taken place. Although the natural gas reserves in the early 1970s equaled only half the energy content of the oil reserves, today the proven reserves for the two fuels are about equivalent. This development can be explained by the fact that the addition of reserves has exceeded the consumption of gas, whereas the situation has been the reverse for oil. It is highly probable that this shift in the resource base in favor of gas will lead to increased use of this fuel in the next century.

This chapter will briefly review the supply and demand situation for gas in Europe and identify some of the more important issues that will determine whether or not a revitalization of gas will come about in the European energy market.

THE NATURAL GAS RESOURCE BASE AVAILABLE TO EUROPE

The four major sources of gas available to Europe (Figure 1) are the gas resources in the Soviet Union, Algeria, Norway, and the Netherlands. It was the large gas reserves discovered in the Netherlands in the early 1960s that initiated the Western European gas development. The gas resource base in the Netherlands is, however, being exploited at a rate that will substantially reduce its importance to the Western European gas supply after the turn of the century.

Major gas discoveries during the past 10 years on the Norwegian

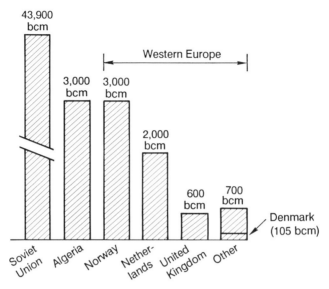

FIGURE 1 Proven reserves of natural gas in the Soviet Union, Algeria, and Western Europe, in billions of cubic meters (bcm).

Continental Shelf have established Norway as an important supplier of gas to Western Europe. The proven and probable gas reserves on the Norwegian Shelf are likely to ensure this role for Norway far into the next century. About half of the Western European proven gas reserves, or 3,000 × $10^9 m^3$ (billion cubic meters), are located in Norway, and the potential resources could be as much as 2 to 3 times larger. Thus, the ultimate gas resource base in Norway could be ranging from 6,000 × $10^9 m^3$ to 9,000 × $10^9 m^3$ or from 212 to 318 trillion cubic feet (Tcf).

The dominant suppliers of nonindigenous gas to Western Europe are the USSR and Algeria. The gas reserves and resources of the USSR are so large that, if unhindered by political forces and considerations about the security of supply, the USSR could potentially supply the total Western European demand for gas far into the next century. This is, however, a highly improbable scenario.

WESTERN EUROPEAN GAS CONSUMPTION

Western European gas consumption has increased substantially during the past 20 years (Figure 2). Of a total energy consumption of 810 × 10^6 tons of oil equivalent (toe) in 1965, gas accounted for 2.5 percent, while in 1985 gas accounted for 16 percent of a total energy consumption of 1,245 × 10^6 toe. It is probable that the availability, public acceptance,

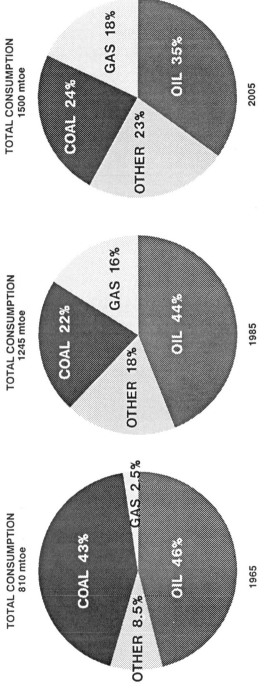

FIGURE 2 Western European energy consumption, 1965–2005, in million tons of oil equivalent (mtoe).

and economics of gas in relation to other fuels will further increase its percentage share of the Western European energy consumption to at least 18 percent in 2005.

Figure 3 shows the gas consumption in the various Western European countries. As will be noted, the penetration of gas in the different markets varies widely. There is considerable scope for increased gas penetration in many of the countries. In particular it can be expected that gas will assume significantly greater shares of the markets in Denmark, Finland, Norway, Sweden, Spain, and Portugal, that is, the countries in the northern and southern periphery of Western Europe.

Since the oil crises of 1974 and 1979, policymakers in Europe have been advised that gas consumption should be promoted where it substitutes for oil, but that gas is too valuable to burn under boilers in industry or for electricity generation. An explicit policy against using gas to generate electricity was laid down by the European Commission in 1982. This policy is now under review in the light of delays in the nuclear program, opposition to coal burning because of the fears of acid rain and the greenhouse effect, and the possibility that the Norwegian Continental Shelf as well as the USSR may offer much greater volumes of gas at competitive prices than has been assumed in the past. With a relaxation of tension between Western and Eastern Europe, there will be less political inhibition about increasing imports from the USSR.

At the same time in certain European countries, the fashion for large electric power units centrally owned and operated is giving way to a belief in small, dispersed local units, for which new gas technologies are becoming available, such as systems that use the waste heat from power generation for district heating. This trend will very likely be reinforced by the privatization of electricity now taking place in the United Kingdom.

Gas for electricity production is also likely to become important in the Scandinavian countries. In Sweden this development will be spurred by the need to substitute new and publicly acceptable power generating capacity for the nuclear power stations that are to be phased out in the beginning of the next century as a result of the political decision to abandon nuclear power by year 2010. In Norway the rising cost of hydroelectric facilities will make gas the least costly way of producing new electricity. In Denmark, for environmental reasons, gas will very likely supplement coal to fuel new electric generating capacity.

A similar development is likely to take place in a number of other countries in the world where discoveries of significant amounts of gas have been made in the past 10 to 15 years.

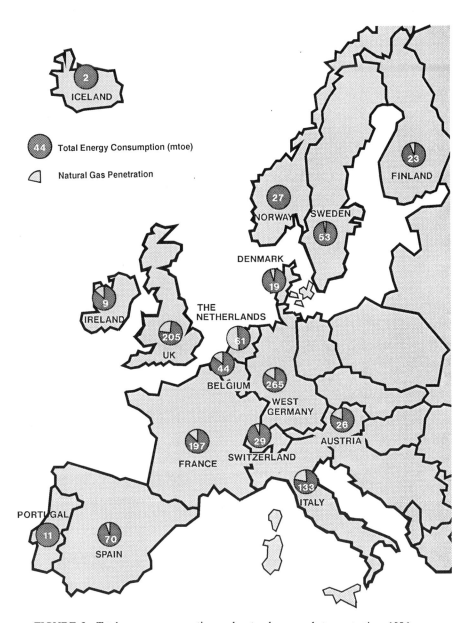

FIGURE 3 Total energy consumption and natural gas market penetration, 1986.

FIGURE 4 Main gas transmission system in Western Europe. The system will be extended as shown by the dashed lines.

WESTERN EUROPEAN GAS SUPPLY

The main gas transmission system in Western Europe is shown in Figure 4. The market for imported gas in continental Europe and the United Kingdom can be reached by a system of trunk pipelines and through liquefied natural gas (LNG) terminals located in France, Belgium, Italy, and Spain. Norwegian gas is available to continental Europe and the United Kingdom through a system of submarine pipelines that terminate, respectively, in Emden, Federal Republic of Germany, and St. Fergus, Scotland. The total capacity of these pipelines is currently about 35 × $10^9 m^3$ per year, or about 3,384 million cubic feet per day (mcf/d). This capacity will be extended to about 50 × $10^9 m^3$ per year, or 4,834 mcf/d by 1995, when a new submarine pipeline, Zeepipe, will connect the Troll Field in the North Sea with the main gas transmission system in continental Europe. This line will have a terminus in Zeebrügge, Belgium.

Although the Norwegian gas reserves have the advantage of being closer to the market than the gas resources in USSR and Algeria, the disadvantage is their high cost of development. A substantial part of the gas on the Norwegian Continental Shelf is located in waters more than 300 meters, or 1,000 feet, deep. This is true of the Troll Field, for example, with recoverable reserves exceeding 1,200 × $10^9 m^3$, or 42 Tcf. Considering the challenges associated with building and operating production platforms in one of the most hostile climatic environments in the world, the high cost of these developments may be understood. Through successive phases of technological development (Figure 5), the industry has, however, gradually learned to build and operate the giant platform structures necessary to cope with these challenges and with predictable unit costs.

In mid-1986, Norway entered into a contract to supply large quantities of gas for a 25-year period (1993–2018) to Germany, France, Belgium, and the Netherlands from Troll and other Norwegian North Sea gas fields. This contract demonstrates the belief of both the sellers and the buyers that

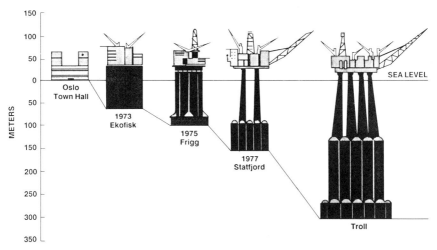

FIGURE 5 Progress in building and operating offshore production platforms at greater depths in the North Sea.

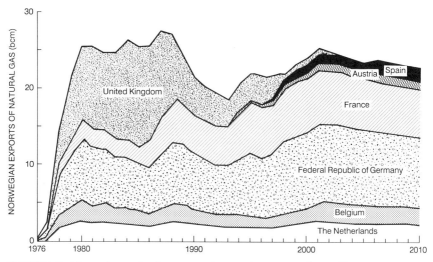

FIGURE 6 Norwegian natural gas export commitments to date. New contracts are expected to increase export volumes substantially in the future.

even this relatively high-cost offshore gas will remain a viable proposition for a long time.

The amount of gas Norway exports on the basis of the contracts already in force is shown in Figure 6. It must be expected that new contracts will substantially increase the export volumes after the turn of the century. In particular, a revitalization of gas exports to the United Kingdom and the development of a substantial market in Scandinavia are highly probable. The necessary pipelines to the United Kingdom are already in place. The pipelines to service the emerging Scandinavian gas market must, however, be developed. Some alternatives for this purpose are depicted in Figure 7.

The technical costs associated with the development, liquefaction, and transportation of gas to the United States from gas fields in the Tromsoflaket (see Figure 7) in the extreme north of the Norwegian Continental Shelf have been assessed. For a sufficiently large scale LNG operation (i.e., 6–10 $\times$ $10^9 m^3$ per year), Norwegian gas could possibly become competitive in the U.S. market around the turn of the century. The key to viable and successful exploitation of the large Norwegian gas resource base is the ongoing research and development to reduce the unit cost of production. For the gas fields in the northern part of the Norwegian Continental Shelf, Figure 8 depicts a unique method for liquefying the gas offshore. Liquefaction of the gas occurs through heat exchange with liquefied nitrogen brought to the field by LNG tankers especially designed for this purpose.

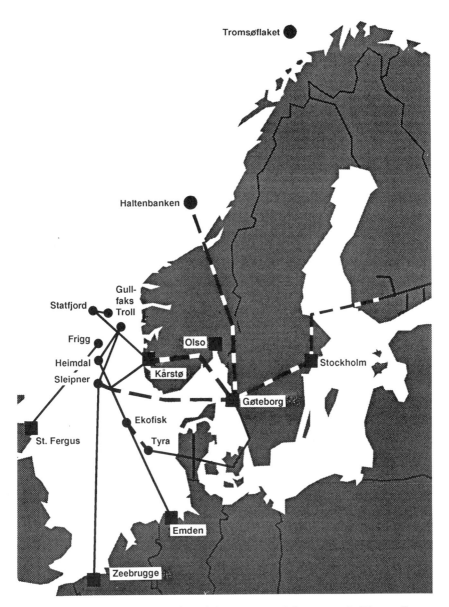

FIGURE 7 Possible additions to the existing gas transmission system in Western Europe to service the emerging Scandinavian gas market.

FIGURE 8 Advanced tanker concept to produce liquefied natural gas (LNG) onboard through heat exchange with liquefied nitrogen (LIN) obtained onshore.

This method of offshore liquefaction has a substantial potential for reducing unit cost.

CONCLUSIONS

Considering the availability of a relatively large resource base for gas in Western Europe and the presence of external suppliers with large gas resources competing for import market shares, it is likely that gas will remain competitive with other fuels and provide the basis for increased gas consumption. The scope for increased gas consumption is further enhanced by the lack of public acceptance for nuclear power in many Western European countries, and stringent and costly regulatory requirements on coal-fired power stations. Thus, gas is likely to reenter as an important fuel in the electricity producing sector.

The technical unit costs for exploitation of the indigenous Western

European gas are tending to increase as a result of high development and transport costs associated with a substantial part of the remaining resources. The research and development effort to cope with these challenges indicates, however, that it should be possible to lower the cost of developing the indigenous resource base, making it competitive with other available and acceptable fuels.

In summary, the technical and economic prospects for increasing Western European gas consumption are good. Western Europe can further develop this increased reliance on gas in its energy balance without becoming overly dependent on nonindigenous gas supplies with the inherent negative political implications. The large gas resources on the Norwegian Continental Shelf thus provide the necessary insurance for a revitalization of gas in the European energy market.

Future Consequences of Nuclear Nonpolicy

RICHARD E. BALZHISER

Despite continuing concerns about the urban environment and growing concerns about global environmental issues arising in part from carbon dioxide (CO_2) discharged to the atmosphere, U.S. energy policy remains strangely silent on the need for options other than fossil fuels.

At the same time, today's market conditions for oil and gas seem to be taken for granted. The presumption seems to be that these fuels will be available indefinitely at current prices without constraint on use—and that coal will play an increasingly important role in electricity supply, with clean coal technologies alleviating urban and regional environmental concerns. Although the latter assumption may be reasonable, the increased use of coal will add to industrial releases of CO_2, which could further accelerate global warming. Elimination of CO_2 from coal plant emissions, although technically feasible, would significantly increase the cost of coal-generated electricity and will give rise to a new waste management problem.

Since the 1960s, the number of environmental laws has increased exponentially, as shown in Figure 1, and we have seen resultant improvements in the environment. We still have a way to go, however, particularly in reducing emissions in the urban areas. Of course, we have paid a price for that improvement, so pollution control costs have also increased exponentially, contributing to increasing cost and declining efficiency for electricity generation after decades of cost decreases (Figure 2). Despite this upward cost trend, the electrification of the nation continues. Since the upward trend started in the mid-1970s, electricity's share of the nation's total energy supply has grown from 28.8 percent to 36.3 percent.

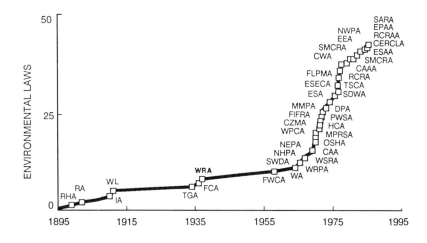

FIGURE 1 Exponential growth of U.S. laws on environmental protection.

Considerable progress has been made at the Electric Power Research Institute (EPRI) in the past 15 years of studying and improving on various alternative methods of electric power generation. Development of clean burning coal technology has been of key importance. Coal gasification has been demonstrated in the highly successful Cool Water project in the Southern California Edison system, and fluidized bed combustion is now being commercialized at utility scale and operating conditions. We may well have a breakthrough in solar electricity, with testing and manufacturing development under way on the highest efficiency photovoltaic conversion device in the world. But despite this progress, we see nothing that could

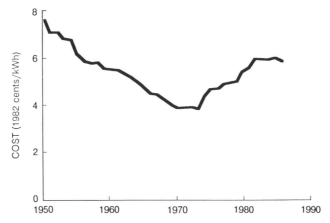

FIGURE 2 Trends in average U.S. electricity cost (1982 cents per kilowatt-hour).

compete with nuclear energy in the long run as it has been deployed in countries such as Sweden, Finland, France, and Japan.

Granted, nuclear power has fallen far short of the euphoric prediction of the postwar years. But it has achieved performance levels, in many U.S. utilities, on the same level of excellence as in France and Japan. What differentiates the successes from the failures in this arena lies not in the technical factors, but in the human and institutional ones. Yet, it appears that the technology has been virtually written off in this country as a candidate for meeting future energy needs. It is my contention that, because of continued environmental concerns—and because of our need for optimum economy and strategic development in the nation's electrical energy supply—we must preserve the nuclear option.

This chapter explores three questions central to the evolution of a comprehensive and rational U.S. nuclear energy policy: What went wrong? How do we fix it? And what are the consequences of continued indifference to the nuclear option?

BUT FIRST, WHAT WENT RIGHT?

Before embarking on this unpleasant but necessary diagnosis of what went wrong, a brief statement is in order as to what went right. Nuclear electric generating capability has been deployed throughout the world at an unprecedented rate, on the order of 5 times faster than any other previous new source of energy. Today 397 nuclear power plants are operating around the world, generating more than 274,000 megawatts of electricity (MWe) in 26 countries (International Atomic Energy Agency [IAEA], 1987). In 1986 alone we saw a 20 percent increase in nuclear power capacity worldwide.

And 133 more plants are under construction worldwide. When completed, these new plants will be producing another 118,000 MWe (Blix, 1987). Another 134 units of 130,000 MWe are now in various planning stages. As of the end of 1986, 4,200 reactor years of operating experience have been achieved (IAEA, 1987).

Many countries vitally depend on the electricity generated by nuclear power. In 1986, as shown in Figure 3, France generated 70 percent of its electricity from nuclear power plants, Belgium 67 percent, Taiwan 44 percent, Korea 44 percent, and Finland 38 percent (Blix, 1987). What may not be as well known is that in the Soviet bloc, Bulgaria generates 30 percent of its electricity from nuclear power, Hungary 26 percent, and Czechoslovakia 21 percent (Blix, 1987). Although the United States is not a leader in percentage, it has the largest total electric output for nuclear power: 85,000 MWe from 108 plants, generating 17 percent of U.S. electric power in 1986 (Blix, 1987).

At the same time, nuclear plant availability is rising. In 1977 the average availability of the 137 units operating around the world was 64.7 percent, comparable if not superior to the existing track record of fossil fueled plants. But in 1986, 288 operating units achieved an average

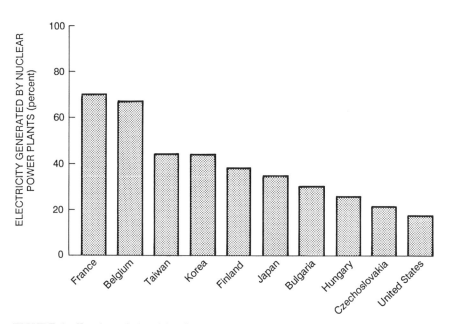

FIGURE 3 Fraction of electricity (in percent) generated from nuclear power in various countries. Although the United States exhibits the smallest fraction, it is the largest producer of nuclear power.

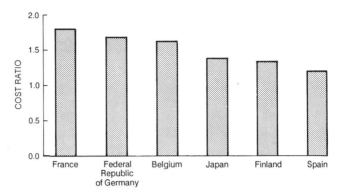

FIGURE 4 Coal-to-nuclear total cost ratio for electricity generation, based on 1986 data. SOURCE: International Atomic Energy Agency (1987).

availability of 70.4 percent, and 55 percent of these achieved a plant availability of 75 percent or better (Blix, 1987). Everything indicates that these impressive improvement trends will continue.

IAEA statistics (IAEA, 1987) for 1986, charted in Figure 4, indicate the following ratios of total baseload power generation costs of coal over nuclear: France 1.8, Federal Republic of Germany 1.68, Belgium 1.62, Japan 1.37, Finland 1.33, and Spain 1.2. The United States is an exception to this trend. On the average, nuclear electricity is not now produced more cheaply than coal electricity, although many individual nuclear plants do so. In fact, as shown in Figure 5, 14 of the 20 U.S. steam electric plants with lowest variable costs (fuel and operating expenses), from 1982 to 1986, were nuclear plants. The nuclear plants recently brought on-line have been so capital-intensive as to slip the total average cost above coal. Nevertheless, the electricity produced by nuclear power in the United States since 1973 has resulted in a cumulative reduction of $65 billion in electricity costs (U.S. Council for Energy Awareness, 1987), mostly due to avoided foreign oil imports. We must not forget that if we set aside amortization costs on the investment, the average variable cost of nuclear electricity in 1986 was less than that of fossil electricity: 19 mills/kWh for nuclear, 21.6 mills/kWh for coal, and 34 mills/kWh for oil (*Energy Daily*, 1987).

An indirect economic benefit has resulted from the substitution of nuclear power for oil to generate electricity. This substitution is estimated to have reduced the world market for oil by as much as $50 billion annually (Lennox and Mills, 1987), thus cutting the pricing powers of the Organization of Petroleum Exporting Countries (OPEC). The resultant lower oil and gas prices have reduced inflation and helped the world recover from the economic recession caused by OPEC.

Add to this the environmental benefits. Under normal operation, nuclear plants are environmentally clean. IAEA data (Blix, 1987) show a 66 percent reduction in sulfur dioxide emissions in Belgium because of increased nuclear penetration into the generation mix, even though a major amount of additional power has been generated. Similarly, reductions of 50 percent in France and 40 percent in Finland have also been achieved. Regrettably, the Chernobyl accident has caused significant land contamination, the consequences of which are still not fully understood, nor is the extent of recovery from the environmental insult.

The safety of nuclear power is a more complex issue. The nuclear industry has compiled an unprecedented safety record among major industrial enterprises. The total industrial safety record of nuclear power is superior to that of alternative methods for generating electricity and of many other activities in our daily lives: the normal annual risk of mortality among the U.S. general population is 330,000 from smoking cigarettes, 120,000 from general accidents, 57,000 from automobile accidents, 7,000 from fires, 6,200 from drowning, 1,100 from accidental electrocutions, 88 from lightning, 10–20 per gigawatt of electricity (GWe) from coal plants, or 2,500, and 0.4 per GWe from nuclear plants, or 35 (Gotchy, 1983).

But concern about the safety of nuclear plants remains high if a severe accident occurs. Chernobyl marks the world's most serious accident involving nuclear power, and it occurred in a reactor system that does not

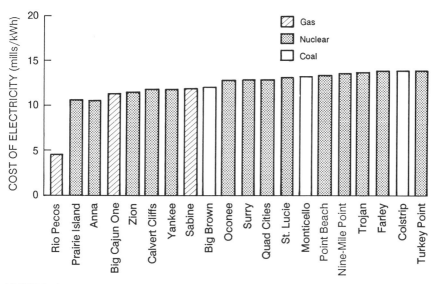

FIGURE 5 Top 20 U.S. steam electric plants with lowest average variable costs over the period 1982–1986.

meet Western safety standards because of its unstable characteristics and its lack of containment. The immediate death toll was 31 plant operators and firemen. Some members of the public will have their lives shortened from cancer induced by the accident. Conservative estimates of incremental cancer incidence are in the thousands. Yet, except in the population that had to be evacuated from the area around the plant, the percentage of increased cancer incidence is so low that the experts judge the increase will not be quantitatively verifiable (U.S. Department of Energy [DOE], 1987).

The nuclear accident at Three Mile Island (TMI), which involved a core meltdown, caused no immediate deaths. The incremental latent cancer incidence is estimated to be between 0 and 1, using the same conservative methods as applied in the Soviet estimates. The importance and effectiveness of containment were demonstrated in that accident. The radiation released in that accident was less than that emitted in the volcanic material from the 1980 eruptions of Mount Saint Helens.

This comparatively good safety record gives no cause for complacence. The record to date does not in itself ensure continuance of a good record. The two severe accidents that have occurred, although not as serious in their consequences as had been expected by the technologists, were financial disasters and have had a profound, negative public and political impact. Thus, there is a need to continue to address the technical safety issues associated with prevention and mitigation of severe accidents.

WHAT WENT WRONG?

We do not have to leave the shores of the United States to diagnose what went wrong with the generally impressive record of accomplishment in nuclear power. There have been two fundamental failures in the performance of the U.S. program. The first failure was the accident at Three Mile Island. Not only did this accident increase public apprehension about the risks inherent in nuclear power, it also reduced the confidence of utility management and the financial community because of the nearly catastrophic financial consequences of the accident for General Public Utilities, the owner of the plant, and for all other nuclear utilities whose credit ratings dropped in the aftermath.

The second failure was in the performance of financial investments in the U.S. nuclear program. Economic problems arose in many individual projects—including significant increases in construction costs and increased unreliability in plant operations, which escalated operating costs. Strong contributors to these problems were the regulator and the intervenor, who, through the judicial system, effected major delays in construction and caused lengthy shutdowns, thereby exacerbating the cost problem. The Calvert Cliffs court decision marked the first time a detailed environmental

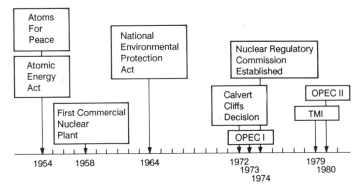

FIGURE 6 Key events shaping the course of nuclear power in the United States.

analysis was required for a nuclear power project, as required by the National Environmental Policy Act (see Figure 6). This decision thus set the stage for major regulatory change and significant delay in nuclear plant construction.

The problem was not helped by the unprecedented high rate of introduction of this new energy technology into a competitive environment with multiple owners and suppliers. This rapid expansion rate, shown in Figure 7, put tremendous strains on the utility industry in implementing the construction and operations programs, did not permit sufficient time to ascend the learning curve, and did not permit an orderly approach to standardization.

The economic consequence of the rapid expansion rate was that the

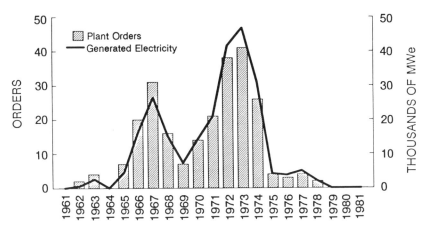

FIGURE 7 Rate of introduction of nuclear power in the United States by generating capacity and number of plants ordered.

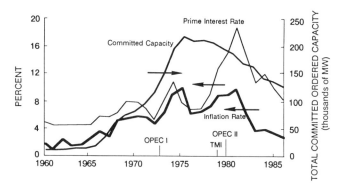

FIGURE 8 Trends in financial environment affecting nuclear power in the United States.

commitment to nuclear power plants reached a peak as inflation and interest rates were reaching double-digit levels, as shown in Figure 8. The delays already resulting from intervention and order backlogs were substantially increased for all plants after the TMI accident in 1979 at a time when prime interest rates approached 20 percent. A capital-intensive technology encountered both high inflation and high interest rates—coupled with regulatory uncertainty and large increases in the time required to complete and license plants—and created the absolute worst-case cost environment. Plant cost increased dramatically, undermining the long-promised economic advantages and providing market resistance to nuclear power. In addition, a large number of backfits and redesigns imposed on the industry by the regulators after the TMI accident substantially increased the direct construction costs and reduced labor productivity.

It is ironic that the seeds of these failures were planted at the outset of the program by two visionary actions: the Atomic Energy Act and the decision to provide containment.

- The Atomic Energy Act was framed to provide a continuing and in-depth level of public participation in the commercial deployment of nuclear power. But this ultimately allowed opposition groups to intervene in nuclear plant construction and operation in any locality.
- The containment design process required studies of the loss of integrity in the reactor systems, which in turn led to studies of the loss of integrity of the containment systems so that the ultimate public risks could be quantified. Such quantification, necessarily at the upper limit, added to the apprehension concerning nuclear power safety; and the need to provide a technically competent, independent regulatory agency to oversee these complex matters became apparent. The Nuclear Regulatory Commission has evolved from this need into its present staff of thousands.

In sum, the decisions to give unprecedented attention to safety and public scrutiny have generated a much greater level of detailed regulatory oversight and knowledgeable public opposition than would otherwise have occurred. There was a blind spot, however, in the implementation of these visionary safety-motivated policy decisions. The focus of development and attention was placed primarily on the nuclear systems. It was assumed that the balance of plant technology, construction methods, operation standards, and maintenance approaches had already been developed in the power industry and could be put to use essentially as is in nuclear power plants. As history shows, this was not the case.

HOW DO WE FIX IT?

The first step in resolving the problems that now beset the U.S. nuclear power program is very much the responsibility of the nuclear power industry. Confidence must be restored in the public mind, in political and financial communities, and among senior utility management—confidence in the safety and effectiveness of nuclear power in the United States. It is imperative that we achieve a uniform level of excellence across the industry with an extremely low incidence of technical problems even under continued detailed scrutiny by the public and the media. And the industry must accept—fair or unfair—the fact that its overall performance will be measured by that of the poorest individual performer. This excellence must also reflect itself in improved economic performance. There is a vital need to turn around the continuing increase in nuclear power operating costs shown in Figure 9.

The U.S. utility industry is fully committed to this drive for excellence and has established three cornerstones to assist in achieving this goal:

1. Formation of the Nuclear Utility Management and Human Resources Council to conduct an integrated review and development of management and people-related issues of nuclear power plant operation in consultation with the commissioners and staff of the Nuclear Regulatory Commission;

2. Formation of the Institute of Nuclear Power Operations (INPO) to establish and monitor high standards of operational performance; and

3. Continued funding of EPRI to provide improved technology.

An important part of these industrial activities is exhaustive scrutiny of plant safety: both enhancing design and operations to minimize the chance of a severe accident, and evaluating and improving accident mitigation features and containment systems to ensure protection of the public in the event of a severe accident. A dominant element in the pursuit of enhanced plant safety is the improvement of human factors in the control room along

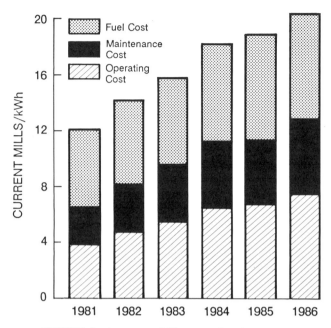

FIGURE 9 Average variable costs of nuclear power.

with the intense effort made by the utilities and INPO—with technical input from EPRI—to increase the level of proficiency in operations and maintenance.

Of equal importance to this operational drive for excellence is the licensing, construction, and operation of a permanent repository for spent fuel. Congress has acted forthrightly to provide for this high-level radioactive waste repository as well as for regional repositories for low- and intermediate-level radioactive waste. The U.S. Department of Energy is dedicated to accomplishing this difficult task.

The completion of the TMI cleanup is another critical remedial action, now in its final phases. At the same time, the utilities have gained tremendous research and development (R&D) value from the cleanup effort. Decontamination methods, robotics developments, equipment qualification diagnoses, and waste-handling methodologies are of value to all nuclear utilities. Unique data are being obtained on the course of a core-melt accident, providing the single most important benchmark of our analytical ability to evaluate severe accidents and their risk to the public. The industry has benefited from the participation and funding of both the Department of Energy and the Nuclear Regulatory Commission in obtaining some of these significant results, supplementing the funding provided by the utilities and EPRI.

The progress being made in this drive for excellence is encouraging. The challenging performance goals set by INPO and the utilities are generally being met, and the success has been noted by several Nuclear Regulatory Commission sources. In testimony before Congress, Lando Zech, chairman of the Nuclear Regulatory Commission, said, "There has been a substantial improvement in nuclear power plant performance over the 1984–1987 period" (Zech, 1988). According to the Office of Analysis and Evaluation of Operational Data at the Nuclear Regulatory Commission, "Overall performance at nuclear power plants steadily improved during 1987, continuing the trend which has developed over the past several years" (Nuclear Regulatory Commission, 1988). Success in these important efforts must continue as a first and fundamental step to a nuclear power future.

In light of the Chernobyl and TMI experiences, an improved international understanding and consensus on nuclear power safety must be achieved, because a nuclear plant accident anywhere in the world affects public acceptance of nuclear plants everywhere. Encouraging steps have been taken. With the recent development of the World Association of Nuclear Operators, all countries with commercial nuclear programs participate freely in an exchange of information on a working level. Moreover, the USSR, a participant in this operator-to-operator exchange, extended a formal invitation to chief executives at nuclear utilities throughout the world to take part in an open forum in Moscow in May 1989.

Even as these measures are taken to restore confidence in present nuclear power generation, there is a need for parallel effort to prepare to reopen the nuclear option in the future. Over the past several years, EPRI and DOE have sponsored a major program to develop an advanced light water reactor (ALWR) for the next increment of nuclear power generating capacity. There is exceptional promise in providing a design that both builds on the extensive operating experience with current designs of light water reactors and incorporates state-of-the-art technological improvement. Utilities in Japan, the Netherlands, Taiwan, and the Republic of Korea are contributing financially and technically to this effort.

For the ALWR to be a viable candidate for utility investment, it must have the following three attributes:

1. It must meet the highest standards of safety and environmental protection.

2. It must be economically attractive in relation to its alternative fossil-fired units.

3. It must provide the utility with a reasonable opportunity to earn a fair return on investment. It must offer predictable construction costs and schedules, assured licensability, predictable operating and maintenance costs, and a near-zero risk of a severe accident. In short, the investor must have high confidence that the large capital investment in the nuclear plant is warranted, and that the investment will sustain its economic superiority throughout the life of the plant.

To achieve these fundamental acceptance criteria, the utility sponsors have established design principles to govern ALWR development, with emphasis on passive safety, simplicity, design margin, human factors, and standardization. Passive safety means that the design principles of the reactor are such that no active safety system is required in the event of a major subsystem failure. These principles are being applied by defining detailed utility requirements for future light water reactor plants, ensuring that the extensive experience gained to date is fully incorporated. Both 1300-MWe and 600-MWe ALWR designs are being sponsored.

Both designs are achieving levels of severe-accident prevention 10 times better than present systems, and both have robust containments incorporating the experience from TMI and extensive testing of containment integrity. Much of the increased prevention capability comes from improved human factors and increased passive safety features. The 600-MWe plant designs have taken a further step in passive safety by incorporating passive decay heat removal features.

The liquid metal reactor (LMR) and the modular high-temperature gas-cooled reactor (MHTGR) have been mentioned as successors to the light water reactor in the United States. In addition to electricity production, each of these systems has its own unique function in the long-term energy strategy of the United States—the LMR as a source of fuel supplies and the MHTGR as a source of industrial process heat. However, these concepts are less mature than the ALWR and will require considerably more development time and the construction of demonstration reactors before they will be ready for commercial introduction to this nation's power grid.

Having discussed the necessary technical fixes, we now must discuss the more difficult and important subject of institutional repair. The needs for regulatory stabilization, Price-Anderson extension,[1] and domestic enrichment supply remedies are well known, and supportive congressional action is being taken in each of these areas.

But there is little action to remove the most formidable barrier to reopening the nuclear option in the United States—the problem of financing a nuclear power plant. Both the regulatory and the financial communities are loath to support any move by a utility to raise funds for nuclear power plants. At the same time, it is doubtful that a utility executive would wish to proceed with construction of a nuclear plant considering the billion-dollar prudency issues that could be placed at his doorstep if the predicted demand does not materialize.[2]

Today's nuclear plant owners face economic risk far greater than anything contemplated before. Their risk goes well beyond the plant itself, as TMI and Chernobyl have dramatically illustrated. Clearly we must develop a symmetry between risks and rewards if we are to reestablish the incentive to build. Either we lower the risk or we raise the potential reward. Today's utility executive is not likely to see either alternative as highly probable in an economically regulated environment, at least not one based on historical rate-making practices. That is not to say that business opportunities could not be structured to offer attractive returns over the lifetime of a project. There is much talk today about restructuring and economic deregulation of the utility industry. It may well be that restructuring and economic deregulation are prerequisites to the authorization of a future nuclear plant. Further study is needed to identify appropriate initiatives for overcoming this barrier.

More important, however, a national consensus on nuclear power is needed to provide a stable foundation for public understanding. The consensus process must be defined. Serious efforts should be brought forth to develop the options for consensus from a broader base of both society and science. There is an important need to introduce new players into this process because of the polarization that has set in among the old players— the industry and the antinuclear segments of the environmental movement. The seemingly irreconcilable positions of these protagonists come from arrogance—an attitude that characterized the industry in the heyday of its unquestioning support, an attitude that has become increasingly apparent in the opposition groups as political and public opinion has shifted in support of them. The players must include the nuclear industry, environmental interests, the ratepayers represented by the financial analysts, nuclear opposition groups, media leaders, and the public as represented by elected officials at both the federal and the state levels.

The approach would consist of a persistent, logical, step-by-step building process. It must be agreed that the issue is important and urgent. Common ground must be defined among all parties with a stake in the outcome. The individual interests must be established and options for reconciling them must be explored. The R&D and evaluations involved in consensus making should entail evaluation of comparative risk and participation by the behavioral and communications sciences as well as the physical and economic sciences.

The toughest and most important of all is public acceptance of nuclear power. As essential as the technical and political steps are, it is not clear that they will be sufficient to turn around public opinion sufficiently to restore the nuclear option. The industry-sponsored U.S. Council for Energy Awareness is active in presenting the case for nuclear power to the public, but it cannot be expected to cause a major turnaround in public opinion. We need a consensus process in which all the interested, key players from all sides of the issue participate.

CONSEQUENCES OF CONTINUED INDIFFERENCE

If we fail to restore the viability of the further use of nuclear power in the United States, we risk major losses. We risk

- exacerbating global environmental problems;
- increasing U.S. electricity rates;
- increasing dependence on foreign oil;
- continued loss of influence in international nuclear policy;
- loss of the human and capital infrastructure necessary to design, deploy, and use the nuclear options;
- loss of the opportunity to export reactors, fuels, and engineering services; and
- loss of the ability to influence how other nations acquire and use nuclear technology.

Two broad ramifications of these consequences are of special concern. First is the loss of the infrastructure that would be needed to expand nuclear power again. This loss carries a correlative weakening of the skilled personnel needed to operate and maintain the present nuclear power capability. The lack of incentive for qualified people to enter an industry that has no future is a formidable problem and inevitably will result in lower staff capability, which will militate strongly against the drive for excellence. The lack of this infrastructure with which to rebuild implies that if nuclear power is needed in the United States again, that need will be met from foreign sources. The potential impact on the trade balance is obvious. The picture of a country the size of the United States being dependent on an overseas supply for a vital form of electricity production is disturbing. Compare this possibility with the overreaction expressed in calls for "energy independence" in the immediate aftermath of the OPEC embargo—plausible scenarios border on the incredible.

ISSUES

Ironically, there is no serious issue that has been raised as a problem for nuclear power that does not have its counterpart in a broad segment of industry today. The difference has been that the nuclear power version of the problem has been identified earlier and publicized more fully in the United States than in other industries.

Radioactive waste disposal, still the single most serious problem for the nuclear industry, has now burgeoned into the widespread issue of toxic waste disposal. The political problems we have in dealing with radioactive waste may appear mind-boggling. Indeed, the public is becoming increasingly cautious about the handling of all wastes, as evidenced by the Long Island garbage scow that wandered the oceans for months in the spring of 1987; "NIMBY" (not in my back yard) has now become a household word.

Emergency response plans are deemed essential to protect lives in the event of a severe reactor accident. Although little attention has been given to this issue by other industries, in several instances emergency response to industrial accidents in the United States has been substantially aided by the organization and preparation provided in emergency planning for nuclear plants.

Another continuing problem for the nuclear power industry has been the assessment of the potential increase in cancer incidence from very low levels of radiation. This subject is still fraught with uncertainty because the radiation levels of interest are so low that the effects cannot really be distinguished from other causes of cancer, particularly from the prevailing natural background radiation. This uncertainty has a counterpart in the rising controversy over the carcinogenic effects of low levels of contaminants in food and even the natural carcinogens in food. Other emerging concerns are associated with the effects of low-level nonionizing radiation such as very weak magnetic fields.

Another dimension of regulatory uncertainty is the issue of prudency, that is, conducting post hoc audits to disallow utility costs. As can be seen in Figure 10, this is another escalating problem in the nuclear industry. But prudency audits are now spreading to other generation systems and will probably grow until the extremes perpetrated on the utilities lead to effective reform.

The problems of nuclear power are largely specific forms of broader issues that affect our society overall. In many respects we are well ahead of the rest of industry in solving them. A *Forbes* cover story entitled "Nuclear Follies" concluded with the question: "In the end, the problem may well boil down simply to this: Can a technology as rigorous and as useful as nuclear power find a place in a society as open as the United States?" (Cook, 1985).

It is becoming more and more apparent that this question applies not only to nuclear power but to most, if not all, high-technology industry. Yet, these technologies are vital to our quality of life. The implications of a negative answer are far more serious than losing the nuclear power option.

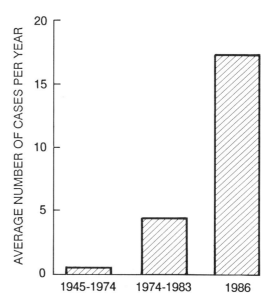

FIGURE 10 Growing trend in prudency audits of utility capital investments.

NOTES

1. In August 1988, Congress authorized a 15-year compromise version of the Price-Anderson Act. The measure raises the pool of potential no-fault insurance funds available to compensate victims of an accident at a nuclear power plant from $700 million to $7 billion. It also exempts Department of Energy contractors from liability for accidents, although it does allow DOE to impose civil penalties on contractors who violate safety regulations. The money would come from two sources: nuclear liability insurance of about $160 million per reactor, which utilities purchase from private insurance companies, and retrospective assessments on each operating reactor in the event of an accident.
2. Such prudency hearings are conducted when a plant is being brought into the rate base to determine whether the utility made a prudent decision in building the plant, given the information available at the time, and whether the utility proceeded with construction of the plant in an efficient manner. If it is determined that the utility may have been imprudent in any of the decisions it took regarding the nuclear project, it may be disallowed to recover a portion of the plant cost through the rate base.

REFERENCES

Blix, H. 1987. Address to International Atomic Energy Agency Conference on Nuclear Power Performance and Safety, September 28, 1987. Vienna, Austria.
Cook, J. 1985. Nuclear follies. Forbes 135:82–100.
Energy Daily. November 12, 1987. Nuclear power seen as least expensive. Washington, D.C.: King Publishing Group.

Gotchy, R. L. 1983. Health risks from the nuclear fuel cycle. In Health Risks of Energy Technologies, C. Travis and E. Etnier, eds. American Association for the Advancement of Science Selected Symposium Series. Boulder, Colo.: Westview Press.
International Atomic Energy Agency (IAEA). 1987. Nuclear Power Status & Trends: 1987 Edition. Vienna, Austria: IAEA.
Lennox, F. H., and M. P. Mills. 1987. An Analysis of the Role of Nuclear Power in Reducing U.S. Oil Imports. Washington, D.C.: Science Concepts, Inc.
Nuclear Regulatory Commission, Office of Analysis and Evaluation of Operational Data. 1988. Report to NRC Commissioners, March 1988. Washington, D.C.: Nuclear Regulatory Commission.
U.S. Council for Energy Awareness (USCEA). 1987. Nader report on nuclear economics. INFOWIRE 87-72. November 6, 1987. Washington, D.C.: USCEA.
U.S. Department of Energy (DOE), Office of Health and Environmental Research. 1987. Health and Environmental Consequences of Chernobyl Nuclear Power Plant Accident. Report No. DOE/ER-0332. Washington, D.C.: DOE.
Zech, L. 1988. Testimony before House Appropriations Water and Energy Subcommittee, March 1988, Washington, D.C.

4
Implications for Strategy

Energy, Environment, and Development

WILLIAM D. RUCKELSHAUS

We are concerned with the effect of energy production on the environment, an effect that has heretofore been seen as a sort of collision. A good deal has been said about many of these collisions: global warming, acid rain, the varied impacts of nuclear energy, and so on. It may seem as though the energy necessary for the sustenance of humanity cannot be produced without wrecking the environment necessary for human survival, but this is an illusion based on shortsightedness and on the failure of some of our political and economic institutions to respond to the wrongheaded. In fact, responsible environmental policy is the only policy that makes sense *economically* in the long run.

Two kinds of experience have led me to this belief. The first, and more recent, was my tenure as U.S. representative on the World Commission on Environment and Development, a panel chartered by the United Nations; the second was my experience as administrator of the U.S. Environmental Protection Agency, both in its founding period and again between 1983 and 1985. Although past experience is not always an unerring guide to the future, it is the only one we have. Whereas the simple extrapolation of current trends is unwise, it seems clear that if some changes are not made in both the way energy is produced and the way the environment is protected, the future will range from unpleasant to awful for most of the people in the world.

This was, in fact, a major conclusion of the World Commission on Environment and Development, which met from 1984 to 1987, and whose deliberations have strongly influenced my own views on environmental protection. The World Commission's charter was both simple and enormous: to formulate a global agenda for change and to propose long-term environmental strategies for achieving sustainable development by the year 2000 and beyond. It proceeded to do this using a remarkable and unique method, that of holding a series of public hearings in major cities in all regions of the world, and receiving ideas and testimony from thousands of people—scientists, scholars, politicians, and large numbers of concerned citizens.

The report of the World Commission, entitled *Our Common Future* (1987), deals with various aspects of development—population growth, food security, resources, energy, industry, urban problems—as part of a single interrelated problem. Its central finding is that the continued prosperity of the developed world depends on the rapid extension of prosperity to the less developed nations in an environmentally responsible manner and that, therefore, economic development and environmental protection are complementary rather than opposing goals, two sides of the same coin. The World Commission has proposed the concept of *sustainable development* as the new model for economic growth, a model that requires efforts to increase prosperity without the destruction of the environment on which all prosperity ultimately depends. This finding, of course, contrasts sharply with earlier studies recommending limitations on growth as the answer to global environmental deterioration.

Naturally, energy development must play a critical, and perhaps the most difficult, role in the realization of this new model. As a matter of fact, the most contentious of all the commission's deliberations were those concerned with energy, and the panel almost failed to reach consensus because of energy-related issues. Perhaps the timing was unfortunate: When the commission began oil was $25 per barrel; when it ended the price had dropped to $10. Also, somewhere in the middle, the Chernobyl accident occurred.

On the other hand, the world may be running out of "good" periods. It is well known that the quarter of the world's population living in the industrialized world now uses about three-quarters of the world's output of energy. This is obviously going to change as dozens of less developed nations push toward large-scale industrialization. Energy demand is going to grow accordingly, and the critical question for the environment is, by how much? Although the answer is essentially unknowable, some reasoned guesses can be made. The logic used by the World Commission can be described as follows.

In 1980 the world consumed about 10 billion kilowatts of energy. If per capita use remained at the same levels as today, the projected world population by the year 2025—8.2 billion people—would use 14 billion kilowatts. However, if energy consumption became uniform worldwide at the current level of industrial nations, the same population would require 55 billion kilowatts. Neither of these figures is realistic; they merely establish the approximate bounds of the range within which energy futures are likely to fall.

The World Commission examined the environmental effects of energy futures at the high and low ends of this range, 35 billion and 11.2 billion kilowatts. The high-end scenario would involve producing more than one and a half times as much oil, more than three times as much natural gas, and nearly five times as much coal as in 1980. This increase in fossil fuel use implies bringing the equivalent of a new Alaska pipeline into production every two years. Nuclear capacity would have to be increased 30 times over 1980 levels.

The high-energy future would continue to aggravate some disturbing environmental trends, directly via physical effects and indirectly through the economies of developing nations. Direct effects include global warming associated with the carbon dioxide produced by burning fossil fuels, as well as urban industrial air pollution and acidification of the environment from the same cause. They also include the various risks—accidents, waste disposal, and proliferation—attendant on the expansion of nuclear energy.

Indirect effects arise from the continuing dependence of less developed nations on steadily increasing amounts of imported energy and their need to borrow vast sums to keep up with demand. For example, the high-energy use scenario just mentioned would require investments of $130 billion per year in the developing countries alone. This dependence creates a desperate need for foreign exchange, which in developing nations often translates into overuse and destruction of natural resources. For example, over 38,000 square miles of tropical forest are destroyed each year, and a like amount is grossly disrupted. It is impossible to tell how much of this loss is attributable to the need to procure energy directly or to pay for it, but that is probably a substantial part and it will continue to grow.

If energy cannot be purchased abroad, it must come from immediately available sources, and in the undeveloped world this most often means firewood. If present trends continue, by the year 2000 nearly 2.5 billion people will be living in areas that are extremely short of fuel wood. In some cities of the developing world, families may pay one-third to one-half of their income for firewood. The pressure on remaining forests from this sort of economics is easy to imagine. Of course, as forests are depleted, we see not only the familiar damage to habitat and species extinctions, but also a diminution in the very ability of the planet to handle the carbon dioxide produced by burning fossil fuels—a vicious cycle indeed.

A future that includes this kind of damage is by no means foreordained, provided we have the political will and the institutional structure to create sustainable development throughout the world. Foremost among the will and structures will be the public environmental demands of the developed world and the agencies created in response to these demands. I would like to offer my own view of U.S. environmental protection—both past and present—because we must understand its capabilities and deficiencies as a tool for solving the problems being addressed.

Here I am both hopeful and dismayed—hopeful because I know, first hand, how far we have come in changing the national consensus on environmental protection. The Environmental Protection Agency has been in existence for less than 20 years. Virtually all our environmental legislation is a product of that brief time-span. Before that, there was a widespread belief among business and political leadership that environmentalism was a fad and that it would, if taken seriously, wreck U.S. industry. That agreement has been entirely reversed. Most all corporate leadership now accepts some form of environmental protection as a legitimate cost of doing business.

Thus, we nearly all have environmental consciousness now, whereas nearly all of us grew up without it. That is a monumental change and a hopeful sign, because if we could achieve such change, then the even greater changes required to establish sustainable development, in energy and elsewhere, may not be beyond our grasp.

The dismaying part results from the current orientation of our environmental protection efforts. In fairness, this orientation arises out of the history of these efforts, a history that might be called "pollute and cure." That is, environmentalism began in this country, as it did in all the industrially developed nations, as a response to widespread pollution. A structure of command and control regulation was established, first for the most egregious pollution and later for the less obvious types. The theory was that by establishing very high standards and gradually cracking down on allowable emissions and effluents, a point would eventually be reached where virtually no pollution would enter the environment.

Where it was appropriate, this approach worked reasonably well, albeit at colossal cost. It was appropriate, for example, in controlling a relatively small number of mass pollutants from easily identifiable fixed and mobile sources. It was appropriate for repairing badly polluted localities through targeted investment in items such as tall smokestacks and sewage treatment plants.

As time went on, new environmental problems emerged, for which this approach was much less appropriate. Thousands of products were found in daily use which, even at very low levels of exposure, had some probability of causing damage to human health or the environment. It was learned that many of the pollution control systems mandated simply transferred pollution from one environmental medium to another—taking toxic wastes out of the river, for example, and burying the residue on the land.

The structure of environmental law and regulation had also become very complex, as the law chased pollution around wherever it seemed most apparent in any particular year. This complexity has rendered almost impossible an ordered, multimedia approach to controlling pollution, in which some finite national investment in pollution control could be aimed at targets that represented the most significant risks.

Most of the environmental protection resources in this country are now directed, as our laws demand, toward reducing even further what appear to be relatively small risks to human health. Very little of that previous resource is left over for dealing with the immense transboundary and global environmental issues that concerned the World Commission, and ought to concern us now.

A slow, legalistic, and extremely expensive system has been created which is at heart an adversarial system. Environmentalists and their political allies push for tighter and tighter controls. The industrial community and its political allies push for lower control costs. Yet, in principle, neither environmentalists nor industry should have any objection to efficient pollution control. We can no longer afford to stage these elaborate battles over incremental pollution, especially when a much wiser goal would be investment in waste-minimizing productive capacity.

What about the rest of the world? Is there some way for nations to achieve environmental goals without eventually reproducing this wasteful and frustrating pattern? The newly industrialized nations have just started to arrive at the stage where they find pollution intolerable. Once this stage arrives, progress can be quite rapid. On Taiwan, for example, a complete reversal of public opinion with regard to pollution control has occurred over the past two years. Taiwan and South Korea will probably increase their environmental consciousness in the late 1980s, not unlike the United States and Japan did in the 1970s. However, these nations will probably not adopt the legalistic, adversarial pattern found in the United States; the national consensus model used by Japan is more likely.

In any case, these nations are not the chief concern over the next 20 years. We should be much more worried about the less developed nations, which are now getting ready for their leap into industrial life. If they must go through the same "pollute and cure" cycle as both the older and the more recently industrialized nations have, three-quarters of humankind may produce pollution at the levels historically produced by the small fraction of it that was industrialized during the century now coming to an end.

Given the current situation with respect to energy, the question must be asked: Will these nations be able to afford it? Highly polluting machinery is often more wasteful of energy and raw materials than its less polluting counterpart. Given the situation with respect to the global environment, another question must be asked: Will the world be able to afford it?

It seems undeniable that somehow, within the next quarter of a century, the transition must begin to a stable base of minimally polluting energy sources at levels that will allow the development and prosperity of all the societies on the planet. It is unlikely that this will be done well unless the power, prestige, and skill of U.S. environmental institutions, public and private, are shifted away from efforts to "control" progressively smaller increments of toxic pollution and toward the long-term problems of the global environment.

For our purposes, these problems can be posed in the form of a single question: How can the world develop the energy it requires and sustain the health of the environment without which it cannot live? Answers must be sought at three different levels with respect to the future: the immediate, the midrange, and the ultimate. These will be addressed in turn.

The immediate issue is how to continue progress toward a sustainable energy future in the current low-price environment. Conventional accounting works against conservation measures when energy is cheap, although paradoxically it is in such periods that more resources are available to make conservation investments against the inevitable day when the price of energy goes up again. From the viewpoint of public policy, there should be no subsidies for fossil fuel use when prices are this low: that means both the familiar direct subsidies and the more subtle environmental subsidies paid via health, property, or environmental damage. Also, policies that discriminate against renewable energy sources should be eliminated. These include both the fossil fuel subsidies just mentioned and the continuing discrimination against small-scale sources of energy by large energy distributors. Overall, these months of low energy costs must be used as a grace period, in which to marshal our resources and establish the basis—through investment and planning—for a sustainable energy future.

In the midrange, ameliorative steps must be taken against the global and regional energy-related problems. This refers mainly to greenhouse warming and precipitation acidification, both of which are vast in scale and subject to considerable scientific uncertainty. In both problems there are a number of plausible scenarios from which to choose.

Consider the following, however: whatever the scenario, the resources that can be devoted to any environmental problem are finite and we cannot afford to launch major programs against every "problem of the week" or to march off boldly in the wrong direction. On the other hand, windows of opportunity may be slamming shut with every year of delay. We cannot afford paralysis by analysis either.

The way out of this quandary seems to be an approach patterned on the way insurance is bought. We are accustomed to sacrifice some present income in order to protect ourselves and our families against the possibility of disaster. No one now would deny the possibility of disaster from these global problems. The arguments are about probability and timing.

Therefore, investments must be adjusted according to the likely range of probability, as with insurance, but in any case at a scale adequate to make a dent in the problem if a dent can be made. The knowledge gained by actually operating a program is invaluable and cannot be replaced by academic research. Moreover, it sends an important message, that the problem is real, and that we are concerned about it. Consider, for example, how much more would be known about how to handle acid rain and how much better off we would be scientifically (not to mention politically) if a modestly scaled sulfur control program had been launched in 1982.

On the ultimate time scale, the basic thing to keep in mind is that global problems require global solutions. It is now possible for one nation to damage another nation inadvertently through environmental pollution at levels of human suffering and property damage that once were associated only with acts of war.

It, therefore, seems wise to accept such problems as falling broadly within the purview of "national defense" and to start paying the kind of attention such damage would demand if inflicted by hostile troops. The recommendations of the World Commission outline what kind of attention is needed.

On the global impacts of fossil fuels, including greenhouse effects and acidification, the commission recommends a four-part strategy that combines improved monitoring and assessment of the evolving phenomena, increased research to improve knowledge about the origins and effects of these phenomena, development of international agreements on the reduction of greenhouse gases, and adoption of international strategies for minimizing damage from the coming changes in climate and sea level.

On the nuclear front, the World Commission recognized that at present, different nations have different views about the necessity and safety of nuclear power. Yet because of the potential for transboundary effects, it is essential that governments cooperate in the development of a comprehensive set of international agreements covering the technical, health, and environmental aspects of nuclear power. These would include such things as international notification of nuclear accidents or transboundary movement of nuclear materials, as well as codes and standards for operator training, compensation and liability, reactor safety, radiation protection, decontamination, and waste disposal.

Above all, in nearly every one of its recommendations, the World Commission urges a return to multilateral action—global responses to global problems. Without an acceptance of this, if global issues are seen only as some legalistic fray between a polluter and a victim, nothing much will be accomplished. In the United States, for example, responsible and wise action on acid rain has been thwarted by, among other things, the insistence that ratepayers of midwestern utilities bear the entire cost of remedial action. In fact, acid rain is, at the very least, a national problem and it requires a national response.

The developed world and its institutions should play a leading role in formulating the global response, but will they? Global responses are difficult things to organize in representative democracies. It is hard for elected officials to spend many chips on efforts that benefit their home constituency only indirectly, or may have some immediate adverse effects on that constituency, and relate to events farther off in time than the next election. On the other hand, as pointed out earlier, no one could have predicted in 1968 the realization of the environmental agenda 20 years later. So perhaps this scant grace period will not be wasted. Perhaps there will be time to plan for the changes attendant on creating the energy future the environment needs, a future with the necessary energy services, at a fraction of current primary energy consumption.

We will, eventually, have to change, and the longer change is put off, the more desperate, painful, and expensive will the remedies be. It remains to be seen for how long narrow considerations of national sovereignty and short-term interest will keep us from doing what global environmental and economic wisdom requires.

REFERENCE

World Commission on Environment and Development. 1987. Our Common Future. New York: Oxford University Press.

What to Do About CO_2

JOHN L. HELM AND STEPHEN H. SCHNEIDER

The energy that people use predominantly comes from burning fuels containing carbon: coal, oil, gas, and wood. When these fuels are burned, carbon dioxide (CO_2) is released to the atmosphere. CO_2 is called a greenhouse gas because it lets radiant energy into the atmosphere more freely than it lets it out. The effectiveness of this atmospheric heat retention increases with CO_2 concentration.

Climate is known to fluctuate naturally over all scales in space and time, yet there is strengthening evidence of a changing, indeed warming, climate, over the past 100 years. Further, this warming is consistent with the progressively increasing concentration of greenhouse gases, especially CO_2, in the atmosphere. Although scientists do not yet know how much of the current climatic change is natural and how much is due to human activity, there is no question that the burning of fossil fuels is the dominant mode of human CO_2 production. There is also no question that a large, rapid climatic change is likely to have a substantial impact on the environment and society.

The subject of this discussion is what can or should be done about the greenhouse situation in view of the attendant uncertainties? To address this question, first we review briefly the essential climatological context of the greenhouse effect and the uncertainties in our understanding of it. Next we review the range of possible climate futures and their potential societal consequences. This provides a framework in which the spectrum of policy responses can be introduced. Finally several energy policy and technology options are presented.

HUMAN ACTIVITY AND THE ATMOSPHERE

Since the industrial revolution, human activity has manifested itself in three similar, large-scale atmospheric problems that have been debated intensely in the past: (1) the possible reduction of stratospheric ozone, (2) the generation of acid rain, and (3) the climatic change from the greenhouse effect. The latter two are actually very old issues. Acid rain has been known for centuries; it was particularly notorious in London, among other places, where coal fires belched oxides of sulfur, which in turn led to the formation of toxic smog. However, only in the past few decades have the substantial long-term effects of precipitation acidity on forests and lakes been scientifically studied in depth. The possible climatic influence of CO_2 has been known for more than a century.

These three problems have several common features: all are complex and punctuated by large uncertainties; all could be long lasting; all cross state and national boundaries; all may be hard to reverse; all are inadvertent by-products of essential economic activities; and all may take investments of present resources to hedge against the prospect of large future environmental changes. The current understanding among atmospheric scientists is that the depletion of stratospheric ozone is not significantly caused by the energy system. Although acid rain is related to energy use, especially coal use, another chapter in this volume will discuss it (Graedel, in this volume). Because current methods of energy production and use present the largest anthropogenic source of greenhouse materials, we will focus our discussion on this issue.

The Greenhouse Effect

Energy from the sun is the principal source of power for life on earth and for the climate. The flow and storage of energy among the various climatic subsystems—the planetary energy balance—is very complex and not fully understood. Figure 1 depicts a simplified version of these flows, and Figure 2 suggests the complexity of the processes by which they interact. The net effect is that the earth's atmosphere acts as a blanket to retain heat and maintain the earth's average surface temperature of about 15°C. This heat retention is principally due to the action of the particles and gases that give rise to the greenhouse effect.

The greenhouse effect, despite the controversy that continues to surround the term, is actually one of the most well-established theories in the atmospheric sciences. For example, with its very dense carbon dioxide atmosphere, Venus has oven-like temperatures at its surface. Mars, with its very thin carbon dioxide atmosphere, has temperatures comparable to our polar winters. The explanation of the Venus hothouse and the Martian

WHAT TO DO ABOUT CO_2

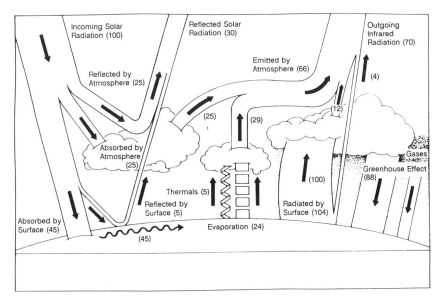

FIGURE 1 Simplified planetary energy balance. The left side of the figure gives the approximate distribution of incoming solar radiation. The right side of the figure shows the flow of terrestrial infrared (IR) energy by water vapor, CO_2, and other gases and particles radiated back toward space. The downward reradiation of IR energy is indicated by the two downward-pointing arrows on the right-hand side of the figure. It is this reradiated energy that provides the basis of the greenhouse effect. Note also that the ratio of energy emitted from the surface layer of the earth (reflected by the surface, 5; emitted by the atmosphere, 25 plus 29; radiated by the surface, 104) is about 1.6 times that of the incoming solar radiation. SOURCE: Schneider (1987a).

deep freeze is quite clear and straightforward: the greenhouse effect (see Kasting et al., 1988). The greenhouse effect arises because some gases and particles in an atmosphere preferentially allow sunlight to filter through to the surface of the planet relative to the amount of radiant energy that the atmosphere allows to escape back up through the atmosphere to space. (The retained energy is represented by the arrow labeled "Greenhouse effect (88)", on the right-hand side of Figure 1.) Note that some level of the greenhouse material is necessary for keeping the planet at a habitable temperature; the earth would be much colder in the absence of any greenhouse warming. Thus, if there is an increase in the amount of greenhouse gases, there is an increase in the planet's average surface temperature, because more heat is trapped (see Figure 3). What is controversial about the greenhouse effect is exactly how much the earth's surface temperature will rise because of an increase in the concentration of a greenhouse gas such as CO_2.

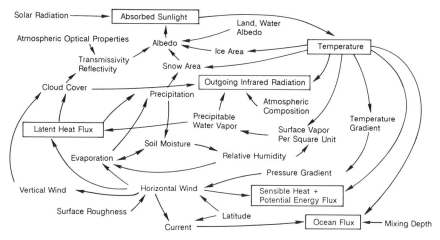

FIGURE 2 Schematic representation of selected important physical processes that affect weather and climate. The interactions indicated by the arrows illustrate several climatic feedback mechanisms. The visual complexity of this figure only partially reflects that of the actual climatic system. SOURCE: Schneider and Londer (1984).

UNDERSTANDING CLIMATE

Although our principal objective is to understand the role of greenhouse gases in the context of climate change, it is helpful to begin by showing how climate changes naturally.

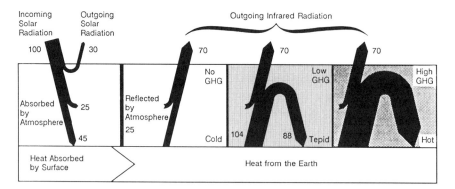

FIGURE 3 Greenhouse gases (GHG) such as CO_2 lead to climate warming because they preferentially retain infrared energy. As the atmospheric concentration of these gases increases, the equilibrium temperature of the lower atmosphere rises to maintain the planetary heat balance.

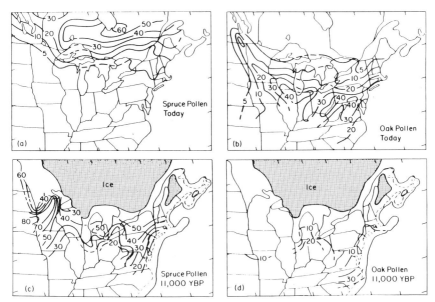

FIGURE 4 Forest pollen distribution since the last ice age more than 11,000 years before the present (BP). Spruce (left column) is better suited to colder climates than oak (right column). As the ice pack slowly receded, the spruce forests moved northward while being replaced by hardwoods in the south. (The numbers are the percentage of fossil pollen accounted for by oak or spruce.) SOURCE: Bernabo and Webb (1977).

Natural Climatic Change

The historical record indicates that the earth's climate has changed and affected ecosystems significantly over geologic time. Figure 4 gives an example of how large, natural, climatic change affects natural ecosystems. This set of panels shows the distribution of fossil pollen found in lake beds and soils over the period since the last ice age. From this and other similar information, it is possible to deduce the spatial distribution of forest species over the past 15,000 years, during which time the ice age conditions gave way to our present interglacial period of warmth. Since the ice age, there has been some 3–5°C global warming, with as much as 10–20°C local warming near where the ice sheets used to be. The spruce species found in the boreal forests in northern Canada today were hugging the rim of the great glacier in the U.S. Northeast and Midwest some 10,000 years ago, whereas the currently abundant hardwood species such as oak were then hanging on in small refugia largely in the South.

The natural rate of forest movement that can be inferred from analyzing the data underlying Figure 4 is approximately 1 kilometer per year, in response to an average temperature change of some 1–2°C per thousand years. If climate were to change much rapidly than this, then the forests might not be in equilibrium with the climate, that is, they may not keep up with the fast change and would go through a period of transient adjustment in which many hard-to-predict changes in species distribution or productivity would very likely occur.

Modeling Climate

To gain a better understanding of climate and the role played by CO_2, a systematic study of the planetary weather system is necessary. Because of the scope and complexity of the planetary energy balance, global climatic theory is not complete; thus, three options for investigation could be considered: conduct experiments, review historical data, or model the system. For obvious reasons it is not possible to conduct a program of global weather experimentation. Close historical analogues are unavailable because there is no period of the earth's history in which the concentration of CO_2 in the atmosphere was, say, twice what it is now, and for which there exists reliable, quantitative knowledge of the climate and ecology. Thus, we now base our estimates on partial natural analogues and climate models (see Schneider, 1987a). These are not laboratory models, since no physical experiments could remotely approach the complexity of the real world. Consequently, we try to simulate the present earth climate by building approximate mathematical models in which known basic physical laws are applied to the atmosphere, oceans, and other elements of the climate system; and the equations that represent these laws are solved with computers.

To model the effects of increasing greenhouse gases we take the best available computer models and change the effective concentration of an atmospheric greenhouse gas. In addition to CO_2, other greenhouse gases include CH_4, N_2O, SO_2, chlorofluorocarbons (CFCs), and tropospheric ozone (Wuebbles and Edmonds, 1988). Taken together, the contribution of these other substances to global warming over the next century is likely to be as important as CO_2 (Ramanathan et al., 1985). However, because CO_2 is presently the predominant greenhouse gas, it has been the subject of most greenhouse simulations. The result of a climate simulation premised on an elevated CO_2 concentration is then compared with a "control" calculation representing the conditions of the earth at present.

The many computer models that have been built over the past few decades are in rough agreement that if CO_2 were to double, then the earth's surface temperature would warm up somewhere between 1.5 and $5.0°C$ (Dickinson, 1986; National Research Council, 1987b). For comparison, the world average surface temperature during the ice age of 18,000 years ago was about 3–5°C colder than our present climate. Thus, a sustained global temperature change of more than a degree or two is a substantial alteration.

Regional Climatic Response

Although the results of current climatic models are useful to scale the magnitude and rate of human alterations to the global climate, they are not sufficient to estimate reliably the societal impacts of climatic change. Rather than focus on the global average temperature, we need to study the regional distribution of evolving patterns of climatic change. Will it be drier in Iowa in 2010, hotter in India, wetter in Africa, more humid in New York, or flooded in Venice?

Unfortunately, to predict the fine-scale regional responses of variables such as temperature and rainfall requires climate models of greater complexity than are currently available. Although preliminary calculations of these variables have been made, it would be hard to reach a consensus among knowledgeable atmospheric scientists about the reliability of regional predictions from current state-of-the-art models (e.g., see Schneider, 1989). Nevertheless, there is at least some suggestion that the following regional changes might occur over the next 50 years:

- Wetter subtropical monsoonal rain belts.
- Longer growing seasons in high latitudes.
- Wetter springtimes in high and midlatitudes.
- Drier midsummer conditions.
- Increased probability of extreme heat waves (see Table 1) and a concomitant reduced probability of extreme cold snaps.
- Forest decline and fire increases in the midlatitudes.
- Increased sea level, by as much as a meter over the next 100 years.

For example, Figure 5 shows the results of a computer simulation of the greenhouse-induced changes in soil moisture associated with drier midsummer, midlatitude conditions. This simulation indicates that such a development could have serious implications for agriculture and water supplies in major grain-producing nations. It must be stressed, however, that considerable uncertainty remains in predicted regional features such as these.

Uncertainties

To consider the uncertainties inherent in current climate models, the topic may be divided into (1) climate model issues and (2) uncertainty about the magnitude of the effective CO_2 source.

TABLE 1 Increased Probability of July Heat Waves due to Climate Warming

Location	Heat Wave Threshold (°F)	Probability Given Current Conditions (percent)	Probability Given 3°F (1.7°C) of Global Warming (percent)
Washington, D.C.	95	18	47
Des Moines, Iowa	95	6	21
Dallas, Texas	100	38	67

NOTE: A heat wave is defined as five or more consecutive days on which the maximum daily temperature exceeds the threshold temperature.

SOURCE: Mearns et al. (1984).

The principal climate model issues involve the crude treatment of hydrological, biological, and other feedback processes in climatic models, and the neglect of the effects of the deep oceans. Feedback mechanisms are best illustrated by an example. If the warming due to added CO_2 were to cause a temperature increase on earth, the warming would most likely melt some of the snow and ice that now exist. Some of the white, highly reflective

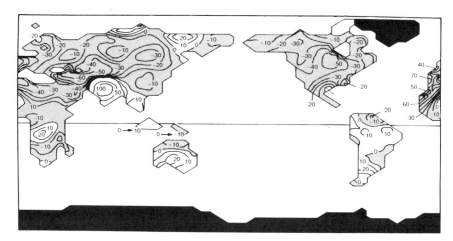

FIGURE 5 Results of a computer simulation of greenhouse-induced change in soil moisture. This change is expressed as a percentage of soil moisture obtained from a control sumulation of present conditions. Change in soil moisture directly affects agriculture and forests. SOURCE: Manabe and Wetherald (1986).

surface of snow and ice would be replaced with darker blue ocean or brown soil, producing surface conditions that would absorb more sunlight. Thus, the initial warming would create a darker planet, which would absorb more energy and thereby accelerate the warming. In this example, climate warming affects surface energy absorption by melting snow and reducing the planet's albedo, which reinforces the warming process further. However, this is only one of several possible feedback mechanisms, and feedback can limit or enhance the process. Because many such feedback mechanisms are interacting simultaneously in the climatic system (see Figure 2), it is extremely difficult to estimate reliably how many degrees of warming the climate will undergo.

The transport of water between various sources and sinks defines the hydrologic cycle. Transformation of water from one phase to another involves the rapid exchange of large quantities of heat. For example, most of the energy in thunderstorms is provided by the heat released as the water condenses from the very humid air rising in the thundercloud. Water, in all its phases, interacts strongly with the other elements of an ecosystem and therefore plays an important role in fine-scale regional climate. A more realistic treatment of the "fast physics" of water-driven cycles is needed to improve the accuracy of regional climate prediction models.

On a global scale, the oceans act as a heat transporter and as enormous thermal buffers, which would respond slowly—over many decades to centuries—to climatic warming at the surface; but they can also act nonuniformly in both space and time. If the atmospheric concentration of greenhouse gases increases as rapidly as typically projected, and if climatic warming were to occur as fast as a few degrees in a century, then the oceans would be out of equilibrium with the atmosphere. Hence, just as we would expect unpredictable transient "adjustments" in forests out of equilibrium with atmospheric conditions, so too we would expect hard-to-predict transient adjustments in an atmosphere out of equilibrium with the oceans.

The natural rate of temperature change during the transition from the last ice age to our present climate was about $1-2°C$ per 1,000 years. The projected rates of change of $2-6°C$ over the next century are more than 10 times faster. Should these rates occur and ocean-atmosphere disequilibrium result, regional forecasts like that of Figure 5 are not very credible. To forecast the global climate under these circumstances, the climate models must realistically include the ocean-atmosphere coupling driven by realistic time evolving scenarios of greenhouse gas increase. This is a formidable scientific and computational task.

Computer climate simulations require as a starting point an estimated value for the effective concentration of CO_2 in the atmosphere. Several interacting biogeochemical processes—called the carbon cycle—control the

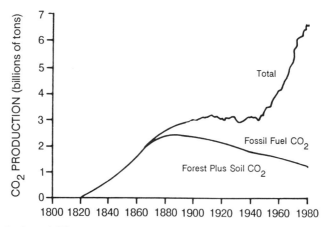

FIGURE 6 Annual CO_2 production, 1800–1980. By the analysis of carbon isotopes in tree rings, a reasonable reconstruction ($\pm$ 20 percent) of the atmospheric CO_2 inventory can be made. Using records of fuel use, the fraction of CO_2 contributed by burning fossil fuels can be determined. This fraction has become dominant since World War II. SOURCE: Bolin (1986).

global accumulation, distribution, and chemical form of carbon. These processes include the uptake of CO_2 by green plants and the slow removal of CO_2 from the atmosphere by biological and chemical processes in the oceans. Uncertainty about CO_2 removal by green plants arises because the rate of photosynthesis is a complex function of climate, the size and distribution of various botanical species, and the health of the ecosystems hosting them. Because the rate of CO_2 capture by the oceans is affected by a complex water-air mixing that occurs at the surface, there is also considerable uncertainty about the rate at which the oceans remove CO_2. For these and other reasons, the fraction of CO_2 that remains in the air is not known very accurately; most current estimates put this fraction in the range of 30 to 60 percent. Although it is difficult to predict what fraction of CO_2 produced today will remain airborne, atmospheric scientists have reconstructed the concentration of atmospheric CO_2 from the past. From such studies, it has been concluded that the use of fossil fuels has now become the dominant source of CO_2, as shown in Figure 6.

CLIMATE FUTURES

How can the effects of climatic change be evaluated? We begin by postulating the likely character of the effects.

Likely Environmental Impacts

One obvious category of direct effects of climate warming is effects on crop yields and water supplies. Another area of concern is the potential for altering the range or numbers of pests that affect plants, or diseases that threaten animals or human health. A wide variety of important effects, such as changes in the probability of catastrophic episodes of drought and flooding, are also possible if climatic change alters the number and character of destructive storms.

Also of interest are the effects on unmanaged ecosystems. For example, the tropical forests are, in a sense, libraries for the bulk of living genetic material on earth. There is major concern among ecologists that, given the present rate of tropical forest destruction through overdevelopment, the world is losing irreplaceable biological resources. The connection between climatic change and this already formidable environmental problem of development and land use becomes clear when one recognizes that substantial future changes to tropical rainfall have been suggested by climate models. These changes imply that current ecological reserves intended to preserve genetic resources may not be as effective as now planned if rapidly evolving climatic change significantly alters conditions in these refugia (Peters, 1989). Simply, they may not sustain even those species that they are designed to protect.

Economic, Social, and Political Consequences

It is certainly easier to postulate the character of the likely climatic effects than it is to assign them a "value." Such an estimation, given a scenario of climatic change, involves more than looking at the total number of dollars lost and gained—were it somehow possible to make such a calculation credibly. It also requires looking at these important equity questions: "Who wins and who loses?" and "Are the losers to be compensated, and if so, how?" For example, if the corn belt in the United States were to "move" north by several hundred kilometers because of climatic warming, then a billion dollars a year lost by Iowa farmers could eventually become Minnesota's billion-dollar gain. Although some macroeconomists viewing this hypothetical problem from the perspective of the United States as a whole might see no net losses in this example, considerable social consternation would be generated by such a climate-induced shift. Such a situation could be exacerbated by the perception that CO_2-producing economic activities were responsible for redirecting the costs and benefits; and these concerns need not be confined to national regions. The perception that the economic activities of one nation could create climatic changes that would be detrimental to another nation has the potential for disrupting international relations—witness our recent experience with acid rain. In essence, the issue of greenhouse gas-induced environmental changes is one of "redistributive justice" (Schneider, 1989).

Risk Assessment

For society to grapple with the environmental consequences of climatic change, a necessary step is to compare various energy options in terms of their climatic risks as well as other collateral social and economic effects. Since the largest single component of the greenhouse effect is the emission of CO_2 from fossil fuel use, it is obvious that any assessment of the potential climatic consequences of fossil fuel use needs to be weighed against risks associated with alternative means of producing and consuming energy.

This is precisely what was attempted more than a decade ago by the Risk/Impact Panel of the National Research Council Committee on Nuclear and Alternative Energy Systems (CONAES). Much of the risk/impact assessment performed at that time is still applicable today and is reproduced here as Tables 2, 3, and 4. Obvious in these tables is the presence of many unquantifiable entries, such as "aesthetic impacts" or "arms proliferation" or "governmental interventions." On the other hand, some risks are more easily quantifiable, such as "accidental deaths and injuries" or "acres of scarred land." The problem for risk analysts, politicians, and citizens alike is how to weigh this complex tangle of entries, each having associated with it some fuzzy, often-intuitive probability. Even for the relatively narrow problem of climatic risks from alternative energy systems, the complexity and uncertainties soon overwhelm the analysis (Schneider, 1979, 1987b).

It is fruitless to hope or wait for a well-quantified assessment of comparative climatic risk assessment across various energy options, let alone a fully comprehensive risk assessment that includes all factors on Tables 2 to 4. Nevertheless, it is possible to quantify many elements of the complex whole. We believe it would be just as foolish to dismiss these well-quantifiable elements as to use them as the sole basis for policymaking. Well-quantified aspects are to be welcomed as helping to place decision making on a firmer factual basis, but they cannot serve as a whole-system risk assessment. There is simply no methodological substitute for human judgment at the whole-system scale.

THE POLICY RESPONSE SPECTRUM

What can or should be done? The greenhouse effect is fraught with scientific and technical uncertainties. It has both potential winners and potential losers. No one nation acting alone can do much to slow it down. Dealing with it substantively could be expensive and could alter life-styles.

TABLE 2 Some Health Risks from Energy Generation and End-Use

Stage of the Fuel Cycle	Nuclear	Fossil Fuels			Hydroelectric	Geothermal	Solar
		Coal	Oil	Gas			
Extraction	Lung cancer Accidental deaths and injuries	Accidental deaths and injuries Lung disease Water contaminants	Accidental deaths and injuries	Accidental deaths and injuries	Accidents in construction	Accidents in construction	Accidents in construction and equipment shipping Toxic by-products of construction materials
Processing	Cancer and mutation from tailings	Lung disease	Lung disease				
Transportation	Theft Accidents	Accidents	Accidents	Death or injury from explosion			
Generation	Cancer and mutation Accidents Sabotage	Bronchitis and exacerbation of chronic disease Possible risk of cancer and mutation Water contaminants Acute respiratory disease; increased mortality Annoyance reactions Altered physiological functions			Accidents from dam failures	Possible toxic effects from emissions (H_2S)	Possible toxic effects (e.g., from metals) Accidents
Transmission	Accidents; possible ozone and electromagnetic radiation changes						
Waste disposal	Cancer and mutation	Cancer, lung disease, mutation				Unknown	
End-use	Deaths and injuries from automobile and other transportation; accidents from other sources; electrocution; fires; accidents from solar rooftop maintenance						

SOURCE: Risk and Impact Panel, National Research Council Committee on Nuclear and Alternative Energy Systems, in Schneider (1979).

TABLE 3 Some Environmental Risks from Energy Generation and End-Use

Stage of the Fuel Cycle	Nuclear	Fossil Fuels			Hydroelectric	Geothermal	Solar
		Coal	Oil	Gas			
Extraction	Scarred land	Scarred land Damaged and decreased water supplies Subsidence Excess rubble	Water pollution from seepage and spills Subsidence	Subsidence	Loss of rivers Silting Estuarine imbalance Water loss	Subsidence Brine discharge Thermal effects H_2S Water pollution Water shortage	Local temperature effects Land use Water shortage Loss of shade
Processing	Mine tailings	Effluents Water shortage	Air and water pollution				
Transportation		Coal dust	Pollution and wildlife harm from spills				
Generation	Thermal effects Water shortage	Acid rain, water pollution particles Climate changes from CO_2; thermal effects; water shortages; NO_x effects					
Transmission	Damage to wildlife, forests, and aesthetics from transmission lines						
Waste disposal	Radioactivity	Excess rubble Trace metals				Pollution	
End-use	Ecological damage from high energy use (photochemical smog) Water shortages Habitat loss						

SOURCE: Risk and Impact Panel, National Research Council Committee on Nuclear and Alternate Energy Systems, in Schneider (1979).

Three classes of actions could be considered. First is countermeasures: taking purposeful actions in the environment to minimize the potential effects (e.g., deliberately spreading dust in the stratosphere to reflect the sunlight to cool the climate as a countermeasure to the inadvertent CO_2 warming). This "geoengineering" solution suffers from the immediate and obvious flaw that if there is admitted uncertainty associated with predicting the inadvertent consequences of human activities, then substantial uncertainty surrounds any deliberate climate modification. It is therefore possible that the inadvertent change might be overestimated by our computer models and the advertent change underestimated, in which case our intervention would be "a cure worse than the disease." The prospect for international tensions resulting from such deliberate environmental modifications is so staggering, and our legal instruments to deal with this possibility so immature, that it is hard to imagine acceptance of any substantial countermeasure strategies in the immediate future. However, countermeasures such as planting more trees to soak up some of the extra CO_2 would be much less controversial. So would removal of CO_2 during energy production by hydrocarbon fuels, were that kind of engineering solution somehow economically feasible.

Second, the policy action that tends to be emphasized by many economists (e.g., Schelling, in this volume) is, simply, adaptation—society should adjust to environmental changes, recognizing that attempts to mitigate or prevent the changes may be prohibitively costly or difficult to implement. We could adapt to climatic change, for example, by planting alternative crop strains that would be more suited to a wide range of plausible climatic futures. Another example of adaption is the building of dikes and other coastal barriers to block the advance of a rising ocean. The U.S. Environmental Protection Agency (Smith and Tirpat, 1988) has estimated that U.S. coastlines can be afforded a substantial reserve of protection against a 1-meter rise in sea level over 100 years by the investment of several hundreds of billions (10^{11}) of dollars; considerable success may be achieved in this regard as witnessed by the Netherlands. The key element of the adaptation argument is that in comparison with the naturally rapid rate at which society rebuilds and changes, the relatively slow rate of predicted climate change should not be very serious.

The third policy category is prevention. In the case of acid rain, this takes such forms as fuel restrictions and use of sulfur scrubbers. In the case of stratospheric ozone, prevention requires abandonment of chlorofluorocarbons and other potential ozone-reducing gases. For CO_2, shifting from coal and oil to gas or reducing the amount of fossil fuel used around the world would prevent some CO_2 formation. Reducing fossil fuel use, often advocated by environmentalists, is controversial because it involves, in some cases, substantial immediate investments as a hedge

TABLE 4 Some Sociopolitical Risks from Energy Generation and End-Use and Comparative Risk Assessment of Electric and Nonelectric Energy Systems

Stage of the Fuel Cycle	ELECTRIC						
				Fossil Electric			
	Nuclear	Solar Thermal	Hydroelectric	Coal	Oil	Gas	Geothermal
Extraction	Land-use impacts Withdrawal from other uses Legal impacts Land rights; access Aesthetic	Land-use impacts Aesthetic Legal impacts Boomtowns Siting impacts	Aesthetic (loss of free-flowing rivers)* Boomtown effects Siting conflicts* Land use* Risks of accidents/ sabotage*	Same as nuclear* (more widespread) Boomtown effects* Water shortages* Reclamation costs* Regional/local inequities	Spill impacts* Aesthetic Recreational		Siting, land-use impacts Land subsidence Noise
Processing	Safeguards* Civil liberties Diversion				Foreign policy impacts (wars, embargoes)* Land subsidence		
Transportation	Safeguards Civil liberties Diversion			Accidents	Refinery siting* Tank farm siting	Pipeline and LNG tankers: availability, safety, conflicts	
Generation	Safeguards* Civil liberties Diversion Siting Fiscal impacts Institutional impacts Risk of accidents*	Siting impacts		Regional equity disputes (mine site vs. remote conversion)* Infrastructure investment			

Transmission	Land-use and aesthetic impacts; power lines System vulnerability		
Waste disposal	Opposition to siting* Long-term care/intergenerational impacts		Capital intensity of control technologies*
End use			
Whole-system risks	Arms proliferation* Political conflicts over safety, economics, equity (local, national, international)* Federal preemption Second-order health impacts	Local political conflicts Equity impacts Federal preemption	Second-order health impacts Conflicts over safety, economics, equity Intergenerational impacts (e.g., CO_2 effects on climate)
	Inflated demand from inefficient energy utilization: greatest dissociation of costs and benefits; regional disputes*		Capital requirements; inflated demand from inefficient energy utilization: greatest dissociation of costs and benefits; regional disputes; CO_2 and climate control*

Continued

TABLE 4 Continued

Stage of the Fuel Cycle	Fossil Direct Use			NONELECTRIC	
	Coal	Oil	Gas	Solar Heating and Cooling	Conservation
Extraction	Same as fossil electric			Legal impacts Sun rights Land-use impacts Scattering	
Processing	Same as fossil electric				
Transportation	Traffic generation				
Generation	Same as fossil electric			High initial costs Operation and maintenance risks	
Transmission				Governmental interventions	
Waste disposal	Capital intensity of control technologies*				
End-use		Transportation impacts*			
Whole-system risks	Second-order health impacts* Regional disputes Allocation problems Intergenerational impacts (e.g., CO_2 effects on climate)			Back-up requirements* Regional imbalances	Capital costs: materials, redesign of industrial processes* Institutional impacts* Curtailment: allocation (equity) problems* Lifestyle impacts* First-cost equity impacts Civil liberties Second-order health risks

SOURCE: Risk and Impact Panel, National Research Council Committee on Nuclear and Alternative Energy Systems, in Schneider (1979).
NOTE: *Means a particularly important risk.

against large future environmental changes, changes that cannot be precisely predicted. Another emission that is hard to control is methane from landfills or rice paddies; methane is a greenhouse gas that could contribute as much as 20 percent of the total heat trapping increase from human activities (Ramanathan et al., 1985).

However, various preventive policies may be effective. The list includes deploying energy-efficient technology, developing alternative energy systems that are not based on fossil fuel, and, in the most far-reaching proposal we have seen, enacting a "law of the air." This was proposed in 1976 by anthropologist Margaret Mead and climatologist William Kellogg (Mead, 1976). They suggest that various nations would be assigned polluting rights to keep CO_2 emissions below some agreed upon global standard (see also Ausubel, 1980; Sand, in this volume; Stavins, 1988).

SUSTAINABLE POLICY RESPONSES: "TIE-IN STRATEGIES"

Clearly, society does not have the resources to hedge against all possible unpleasant futures. However, without anticipatory actions society may well suffer unnecessary and substantial losses during climate change transition. The nature and extent of such transitional losses depend critically on the *rate* at which adaptation must occur. The oil price shocks help to illustrate this point. Had oil prices risen slowly over many years, the United States would have been able to adapt better to the changing price environment. Instead, there was little time for anything but panic and painful adjustments. (However, the price rise did lead to major improvements in energy efficiency that eventually helped bring the price down.)

Since each hedging strategy has an associated cost, how can the optimum level of investment be determined? A complete discussion of this topic is beyond the scope of this chapter; however, one guideline we will discuss here has been called the "tie-in strategy" (Boulding et al., 1980; Kellogg and Schware, 1981), that is, those actions that provide widely agreed upon societal benefits even if the predicted changes do not materialize.

Several examples of tie-in strategies can be identified. The most obvious is to generate, distribute, and use energy as efficiently and cleanly as possible.

Energy efficiency is a tie-in strategy because climatic change is only one of several good reasons to consider a policy of energy efficiency. What would be wasted by an energy efficiency strategy if serious global warming never materialized? Although the rate of investment in efficiency does depend, of course, on other competing uses of those financial resources, it often makes good economic sense to use energy more efficiently. The "tie-in" strategy suggests priority for policy responses to the advent or prospect of the greenhouse effect in ways that would have societal benefits even if present estimates of the severity of the greenhouse effect are not met. If in the equally likely event that current estimates are too small, then the efficiency strategy will have been beneficial because greater efficiency in energy use can reduce undesirable vulnerabilities to foreign sources of energy and reduce other non-CO_2 emissions of pollutants and their undesirable health effects. Moreover, energy efficient manufacturing can in many cases reduce production costs and enhance global competitiveness of U.S. industry.

In addition to making more efficient use of the energy we produce, it is also important to produce and use it cleanly. In principle it is possible to analyze the greenhouse tie-in benefits for an entire energy system. However, to do so is beyond the scope of this chapter because of the wide variety of possible systems. One major basis of such an analysis should be the "net greenhouse gas cost." Although energy systems may produce several possible greenhouse gases, since non-CO_2 greenhouse gases can be expressed in terms of a CO_2 equivalent, it is convenient to introduce a simplified unit called the "net CO_2 cost." The net CO_2 cost is defined as the ratio of total gaseous carbon (e.g., moles or tons of CO_2, CO, or hydrocarbons) emitted to the total *useful* energy produced. The amount of total gaseous carbon can be obtained by a *system lifetime CO_2 balance*. By system lifetime we mean the amount of CO_2 produced as a result of materials fabrication, construction, operation, decommissioning, and disassembly. The meaning of useful energy depends somewhat on how an energy system is defined. For example, if the waste heat from electricity generation is used for industrial process heat or space heating of homes, the useful energy total should include that portion of the heat so used. Employing CO_2 cost as we have defined it here makes possible the comparison of different systems, such as nuclear, hydroelectric, geothermal, ocean thermal, and solar. Then, if two systems are equal in all respects except their CO_2 cost, we would advocate that the system with the lower CO_2 cost should be chosen because it represents a tie-in strategy or perhaps that a fee be charged on CO_2 costs. CO_2 costing has the useful property that a clear definition of the entire life cycle of the system is required before deployment.

CO_2 cost analysis, of course, is only one of many criteria that should be used to evaluate the suitability of an energy system or subsystem. We also caution that because CO_2 costing is based on our imperfect, present understanding of future technology and practices, it, like whole-system risk assessment, should be used more for qualitative guidance rather than as a quantitative performance measure. This is especially true with large-scale infrastructures such as energy systems, because they are sufficiently long-lived for improvements in technology and other pleasant surprises to occur. Note, for example, how technology continuously improved domestic oil production over the past 15 years (Bookout, in this volume).

On a net CO_2 basis, we know that coal typically emits about twice as much CO_2 per unit of energy released as methane. Because it is certain that reduced emissions from fossil fuels—especially coal—will reduce acid rain and the negative health effects of air pollution in crowded areas, we see that another tie-in strategy is to emphasize the use of hydrogen-rich fuels. Specifically, encourage the use of methane and emphatically discourage the use of coal. The tie-in benefits follow from the increased efficiency and reduced pollution made easier by fuels such as methane, provided that uncontrolled release of methane, itself a greenhouse gas, is adequately controlled.

There are many nonfossil-based energy systems that can produce energy without CO_2 emission. Of course, CO_2 is not the only environmental side effect of any of these (see Tables 2 to 4); however, it has certainly become an important one. The most developed of the nonfossil energy technologies is electricity generation by nuclear fission power plants, which emit no CO_2 during actual plant operation and which should be considered as a potential tie-in strategy. To hold to the concept of CO_2 cost, however, the extent to which nuclear power is a tie-in strategy will depend on its total CO_2 cost, not just its operating CO_2 cost.

The increasing prospect of CO_2-induced climate change has added CO_2 generation to the risk assessment equations, and this addition introduces new dimensions of comparison. For example, when comparing nuclear and fossil electricity generators, the question arises as to which poses a more serious risk: enormous volumes of low-activity waste (CO_2) or small volumes of high-activity waste (fission products)? The application of CO_2 costing may help in addressing such questions. Therefore, we advocate that CO_2 cost be evaluated for all important energy technologies, including nuclear electricity generation, and be used as an additional criterion in system selection and evaluation.

The tie-in benefits of efficiency technologies and hydrogen-rich fuels are clear and in many cases could be relatively inexpensive. However, we wish also to call attention to some important, but more embryonic, tie-ins such as greenhouse-mitigating transportation infrastructure, hydrogen as a fuel, and genetically engineered plant species.

As future transportation infrastructure is planned, several greenhouse tie-ins can be identified. For example, the daytime population of cities such as New York would be impossible without the infrastructure of commuter trains and subways. Making these systems even more efficient and desirable means of daily travel is a tie-in strategy that substitutes the use of CO_2-intense means of travel with a more CO_2-benign means of travel. This substitution operates at two levels: (1) The amount of CO_2 emitted per train passenger mile is much less than is emitted per car passenger mile, even if the train is powered by a large fossil energy source (i.e., a large diesel or a fossil-fired powerhouse); (2) Once a modern commuter transit infrastructure is in place, it facilitates the substitution of high-CO_2-emitting power sources by power sources that are environmentally more suitable. Examples include electric trains powered by solar or nuclear-generated electricity, or possibly a hydrogen-fueled locomotive.

Even greater opportunity exists in the so-called corridors between U.S. population centers (Marchetti, 1988), where high-speed train service makes more sense from the standpoint of reducing greenhouse warming. True integration of transport systems is an important component of a train renaissance in these corridors. The airport in Frankfurt, Germany, provides an example of a high level of integration; it is constructed in levels, one for the European train system, one for cars and buses, and one for aircraft. In contrast, most U.S. cities with urban train systems do not provide services that effectively connect with the terminals. However, some progress in urban transportation is occurring in the United States today. For example, the Chicago mass-transit system has a line that ends in the bottom level of a terminal building at the airport, but unfortunately it is a long walk to the aircraft gates. And in Philadelphia, there is now direct train service available from downtown to the terminal at the airport. Even without concern over CO_2 emissions, these transportation improvements are attractive.

Hydrogen may prove to be the ultimate fuel. It is clean, renewable, and can serve as a truly fungible "energy currency" because it can be readily converted into other forms of energy. However, many critics of hydrogen oppose its use on the basis of cost, safety, and the absence of complementary technologies for its transportation and use.

The cost of hydrogen is, of course, a function of feedstock price, processing cost, and demand. Hydrogen can be made in a variety of ways; but most likely the first large quantities of hydrogen will be produced by methane steam reforming, which is projected to produce hydrogen at 1.2 times the price of petroleum-derived liquid fuels on a per unit energy content basis (Scott, 1987). For technologies designed to exploit the advantages of hydrogen, its slightly higher cost as a fuel is more than offset by its form value. Evidence of how form value can offset energy cost is provided by many of the uses of methane and electricity today. For example, the economics of electricity generation achieved with new gas turbine technology more than offsets the higher fuel costs of methane (McCormick, in this volume). As superior technologies for hydrogen production come on-line, hydrogen production should become even more economical; and therefore over the long run, hydrogen may be price competitive with other energy carriers, such as petroleum-based fuels. Although safety problems need to be addressed, they may prove to be more perceptual than actual, as was the case when gasoline was introduced as a motor fuel.

In rebuttal to concerns about the absence of a hydrogen technology, it is not necessary to have a complete analogue of our fossil-fueled transportation system for hydrogen to become important and useful. For example, hydrogen can be a viable fuel for systems such as high-performance aircraft and can function as an energy currency in that it is convertible to other energy forms such as electricity and hydrocarbons. Further, the beginnings of a hydrogen technology are already well developed. For example, hydrogen technology has been central to the U.S. manned space program for decades; large quantities of hydrogen have been generated for ammonia synthesis, hydrocracking of petroleum, hydrotreating of hydrocarbons, and methanol synthesis; and in the Federal Republic of Germany hydrogen producers and users have been safely linked for more than 30 years by a 208-kilometer hydrogen pipeline, portions of which run through several cities. Similar pipelines exist in the United States to transport chemical intermediates among processing plants.

The United States has a long and fruitful history of creating technology where there was none. Therefore, the absence of complementary hydrogen technologies is not a reason to oppose its eventual use. Other opportunities provided by hydrogen have been studied elsewhere (e.g., see Scott, 1987), and may even be materializing; as evidenced, for example, by the recent agreement signed by the European Community and Quebec, Canada, to conduct a five-year study on shipping hydroelectrically generated hydrogen from Canada to Europe for public transportation fuel. The key issue for hydrogen is that its use is environmentally benign and it can be produced from a variety of nonfossil sources such as hydroelectric facilities, photovoltaic devices, or, possibly, inherently safe nuclear plants. Of course, any system for hydrogen production, distribution, and use must be evaluated in terms of its net greenhouse gas cost.

The technology of genetically engineered crop strains is quite new. We only note here that developing crop strains that are climatically more robust, for example, drought resistant and able to use higher CO_2 levels effectively, are desirable for many reasons. Although it would be too speculative to conjecture what might be done, it is easy to see how such crops also make sense as tie-in strategies for CO_2 adaptation or mitigation (National Research Council, 1987a). Other tie-in benefits can be argued for trade agreements or treaties for climatically dependent strategic commodities such as wheat and water.

In some circles there would be ideological opposition to government funding of all such tie-in strategies on the grounds that these activities should be pursued by individual investment decisions through a market economy, not by collective action using tax revenues. In rebuttal, it can be pointed out that exactly this kind of strategic investment is made on the basis of noneconomic criteria even by the most conservative people: investments in military security. It is not an economic calculus that dictates investments in a military, but rather a strategic consciousness. The argument here is simply that strategic consciousness can be extended to other potential threats to our security, including a substantially altered environment occurring at unprecedented rates.

Furthermore, a zero cost of pollution hardly sets up proper market incentives for energy efficiency or alternatives. Recent experiences with pollution controls and hazardous waste cleanup have already shown that pollution is not cost free. Quite simply, the land, water, and air that historically have been cost-free inputs to the metabolism of society are now becoming more expensive to use. The cost of pollution defines an additional economic feedback not unlike the feedbacks in the climate system we discussed. Because these future liabilities are not well quantified, they are not reflected in the current prices of these "free" resources. Investment to hedge against potential environmental change can, however, deny resources to other socially worthy goals. More research will certainly put policymaking on a firmer scientific basis, but credible details about specific winners and losers are not likely to be available much before society has committed itself to large atmospheric changes. If we choose to wait for more certainty before taking preventive actions, then this is done at the risk of having to adapt to a larger, faster-occurring dose of greenhouse gases, acid rain, and ozone depletion than if actions were initiated today.

In sum, many of the prudent things to do about the greenhouse effect are prudent things to do anyway.

REFERENCES

Ausubel, J. H. 1980. Economics in the air. Pp. 12–59 in Climatic Constraints and Human Activities, J. H. Ausubel and A. K. Biswas, eds. Oxford: Pergamon Press.

Bernabo, J. C., and T. Webb III. 1977. Changing patterns in the holocene pollen record of northeastern North America: A mapped summary. Quaternary Research 8:64–96.

Bolin, B. 1986. How much CO_2 will remain in the atmosphere? Chapter 3 in The Greenhouse Effect, Climatic Change, and Ecosystems, B. Bolin, B. R. Döös, J. Jäger, and R. A. Warrick, eds. New York: John Wiley & Sons.

Boulding, E., et al. 1980. Pp. 79–103 in Carbon Dioxide Effects Research and Assessment Program: Workshop on Environmental and Societal Consequences of a Possible CO_2-Induced Climatic Change. Report 009, CONF-7904143. Washington, D.C.: U.S. Government Printing Office.

Dickinson, R. E. 1986. How will climate change? The climate system and modelling of future climate. Chapter 5 (pp. 207–270) in The Greenhouse Effect, Climatic Change, and Ecosystems, B. Bolin, B. R. Döös, J. Jäger, and R. A. Warrick, eds. New York: John Wiley & Sons.

Kasting, J. F., J. B. Pollack, and O. B. Toon. 1988. How climate evolved on the terrestrial planets. Scientific American 258(2):90–97.

Kellogg, W. W., and R. Schware. 1981. Climate Change and Society, Consequences of Increasing Atmospheric Carbon Dioxide. Boulder, Colo.: Westview Press.

Manabe, S., and R. Wetherald. 1986. Reduction in summer soil wetness induced by an increase in carbon dioxide. Science 232:626–627.

Marchetti, C. 1988. Infrastructure for movement: Past and future. Pp. 146–174 in Cities and Their Vital Systems: Infrastructure Past, Present, and Future, J. H. Ausubel and R. Herman, eds. Washington, D.C.: National Academy Press.

Mead, M. 1976. Preface to society and the atmospheric environment. In the Atmosphere: Endangered and Endangering. Fogarty International Center Proceedings No. 39, W. W. Kellogg and M. Mead, eds. Washington, D.C.: Department of Health, Education, and Welfare Publications.

Mearns, L. O., R. W. Katz, and S. H. Schneider. 1984. Extreme high-temperature events: Changes in their probabilities with changes in mean temperature. Journal of Climate and Applied Meteorology 23:1601–1613.

National Research Council. 1987a. Agricultural Biotechnology: Strategies for National Competitiveness. Washington, D.C.: National Academy Press.

National Research Council. 1987b. Current Issues in Atmospheric Change, Summary and Conclusions of a Workshop October 30–31, 1986. Washington, D.C.: National Academy Press.

Peters, R. L. 1989. Proceedings of the Conference on the Consequences of the Greenhouse Effect for Biological Diversity. New Haven, Conn.: Yale University Press.

Ramanathan, V., R. J. Cicerone, H. B. Singh, and J. T. Kiehl. 1985. Trace gas trends and their potential role in climate change. Journal of Geophysical Research 90:5547–5566.

Schneider, S. H. 1979. Comparative risk assessment of energy systems. Energy 4:919–931.

Schneider, S. H. 1987a. Climate modeling. Scientific American 256(5):72–80.

Schneider, S. H. 1987b. Future climatic change and energy system planning: Are risk assessment methods applicable? In Proceedings of the Engineering Foundation Conference on Risk Analysis and Management of Natural and Man-Made Hazards, November 8–13, 1987, Santa Barbara, Calif.

Schneider, S. H. 1989. The greenhouse effect: Science and policy. Science 243:771-781.

Schneider, S. H., and R. Londer. 1984. The Coevolution of Climate and Life. San Francisco, Calif.: Sierra Club Books.

Scott, D. 1987. Hydrogen: National Mission for Canada. Cat. No. M27-6/1987E, Ministry of Supply and Services, Canada.

Smith, J. B., and D. A. Tirpak, eds. 1988. The Potential Effects of Global Climate Change on the United States: Draft Report to Congress, Vols. 1 and 2. U.S. Environmental Protection Agency, Office of Policy, Planning, and Evaluation, Office of Research and Development, Washington, D.C.

Stavins, R. N., ed. 1988. Project 88: Harnessing Market Forces to Protect Our Environment—Initiatives for the New President. A public policy study sponsored by Senator Timothy E. Wirth and Senator John Heinz, U.S. Congress, Washington, D.C.

Wuebbles, D. J., and J. Edmonds. 1988. A Primer on Greenhouse Gases. DOE/NBB-0083. Washington, D.C.: National Technical Information Service.

Achieving Continuing Electrification

WALLACE B. BEHNKE, JR.

A scenario of continuing electrification such as that of Chauncey Starr (in this volume) implies the ongoing availability of adequate amounts of reliable and reasonably priced electric power. Therefore, it seems virtually certain that for at least the next several decades U.S. electric power supply systems will play an increasingly critical role in the nation's energy economy. Rather than debate this point, let us focus on the upstream implications of continuing electrification for the electric power supply infrastructure.

The deliverable supply of energy, including that provided by the electric power industry, will be affected by the interplay of public policy, technological innovation, environmental concerns, and economic forces. As things stand now, U.S. electric power supply systems face the prospect of fundamental change engendered by the energy dislocations of the 1970s and nurtured by the policies and politics of the 1980s. Hard choices will have to be made during the 1990s, and the legacy of those decisions will endure well into the next century.

This chapter focuses on several of the more significant forces that promise to reshape the nation's electric power supply systems over the next several decades, beginning with demand.

DEMAND

Any forecast of electricity growth on a global scale to the year 2060 is a courageous and useful attempt to put the issues into perspective. Is a projected tripling over the next 60 years optimistic or pessimistic?

We simply do not know. We do know that, despite increasing emphasis on conservation and end-use efficiency, improvements in productivity and quality of life have caused the U.S. electric power systems to experience new records for peak demands and output. Is it reasonable to expect any less of the developing nations? Probably not. There is also evidence that the growth of energy use occurs in long pulses. If we are nearing the end of a slow period, will the current trend in public policy, which discourages utility construction in favor of contractual arrangements with unregulated suppliers, prove unexpectedly detrimental?

Despite the uncertainty of demand projections, improvements in end-use efficiency brought about by the combination of new technology and market forces could postpone the need for new generating capacity. Efficiency gain on the customer side of the utility meter is viewed as an objective worth pursuing. Recognizing this, utilities are spending more to promote conservation. Moreover, at least 20 states now impose requirements for least-cost planning, which incorporates both supply- and demand-side strategies. It has even been suggested that utility payments to customers for certain conservation improvements would be more economically efficient than investment in additional utility plants and should be passed through to ratepayers. The merits and ultimate costs of this concept are the subject of intense debate. Even so, experimentation in this direction is already under way.

SUPPLY

The outlook for electricity supply should also be considered. The North American Electric Reliability Council oversees and promotes power system reliability for North America's electric utilities. The council was created by the utilities in response to the system black-outs of the 1960s. In a 10-year forward assessment of bulk electric system reliability, the council found the supply plans of U.S. electric utilities adequate to support an average annual growth in electricity demand of 2.0 percent per year (North American Electric Reliability Council, 1987). Although these are seen as the most likely growth rates, the council estimates a 50 percent chance that growth will be greater than these utility plans can support. Two things are troubling about this assessment:

1. Since 1982 when the United States emerged from the recession, electricity sales growth has been higher than 2 percent annually—growing 3.5 percent per year on average. In 1987, sales were up 4.5 percent, and for the first quarter of 1988 they ran at an annual rate of 5.9 percent above the correspondending rate of 1987, well ahead of the gross national product in 1988.

2. Construction has not yet started on almost 45 percent of the 79,300 megawatts of new generating capacity currently planned in the United States through 1996.

Much of this capacity will not be completed on schedule because of drawn out regulatory proceedings, construction delays, cost containment pressures, and increasingly stringent environmental requirements. The council predicts that supply problems could develop in some areas of the country as soon as 1990. Offsetting concerns about adequate supplies are claims that these projections are too high, especially if more emphasis is given to efficiency opportunities and other demand-side strategies. Which assessment is correct is not known.

RELIABILITY

The chapters by Weinberg, Gibbons and Blair, and Starr in this volume illustrate the profound difficulty in forecasting electricity demand with a sufficient precision to be useful for planning the additional capacity, given the 10- to 12-year lead times currently required to build new central station generating facilities and the useful life they must serve. The blackouts of the mid-1960s demonstrated that the consequences of a breakdown in the power supply system are greater than the costs of having new capacity in service a few years sooner than needed for system reliability. This experience continues to temper the judgments of utility system planners, but it is apparently being largely ignored in the current public policy debate over the power supply issues.

The U.S. bulk power transmission system has evolved into a complex network. Portions of this system will continue to be heavily loaded by energy transfers, both within and among regions, as utilities strive to minimize the cost of electricity. In some areas, concentrations of non-utility power generation will further increase loadings on already heavily used segments of the grid. In addition, energy transfers among regions can increase loadings in utility systems not party to these transactions. Similarly, disturbances in one system can affect the reliability of other systems. All of these factors underscore the critical importance of coordinated planning and operation of the bulk electric transmission systems.

Reinforcements of the bulk power grid are planned to enhance energy transfer capability and ensure continued reliability. About 20,000 circuit miles of new transmission capacity are planned for service over the next 10 years. Yet, lengthy licensing proceedings and other disincentives inherent in the current regulatory process create formidable impediments to constructing new transmission lines. The North American Electric Reliability Council warns that delays encountered in the expansion of the transmission grids will further compound the stress on the power supply network by reducing operating margins and limiting the flexibility to respond to facility outages, start-up delays, fuel supply interventions, and other system emergencies.

DEREGULATION

A second major force shaping the electric utility business in the United States is deregulation. Deregulation has implications for both the price and the reliability of electric service.

After deregulation of the telecommunications, airline, trucking, and natural gas industries, interest in more competition in the electric utility industry has grown. In theory, increased competition will produce lower prices and a more efficient industry.

The bulk power markets will almost certainly become more competitive in the years to come. Increased competition from qualified generating facilities (QFs) and independent power producers (IPPs), encouraged by the Public Utility Regulatory Policy Act of 1978, is already being felt in the bulk power markets and could extend to the retail markets as well. At the same time, a more adversarial regulatory climate has caused utilities to be more averse to risk. Many utility managers say they plan to meet future electricity demand by extending the life of existing facilities, purchasing power from other suppliers, and promoting conservation and load management.

The Federal Energy Regulatory Commission (1988) is responding to this situation by holding hearings on proposed rules that would expand opportunities for IPPs to compete in the bulk power market. The commission apparently believes that IPPs are needed as a new and flexible source of power at a time when utilities are reluctant to invest in new generating capacity. The commission is also considering guidelines for the conduct of competitive bidding as a means of purchasing power from unregulated producers.

In weighing the merits of deregulation, a number of key questions should be addressed. For example, to what extent will deregulation cause utilities to lose control of construction and operation, and what will be the consequences of this loss? The highly reliable electric power systems in North America are the result of extensive planning and operating coordination among utilities over many years. Because of antitrust concerns and competitive conditions, will power producers withhold from each other the very information needed for coordination to maintain reliability? Can centralized, coordinated planning, which some advocates of least-cost planning seek to enhance, coexist with competition? What characteristics must an IPP have for its operation to be fully coordinated with the rest of an electrical power system? What contractual arrangements would be necessary for

IPPs to be willing or able to engage in economic dispatch? How would the quality of service be impacted by a large concentration of IPP generation?

Other important questions also need to be considered. Encouraging regulated utilities to seek bids for new bulk power requirements from QFs and IPPs may help ensure that the most efficient sources of power are developed, provided there is a level playing field for all participants, including the utilities. Can utilities develop the economic characteristics of a competitive enterprise with the continuing obligation to be the supplier of last resort? If utilities are excluded from bidding on capacity to serve their native load as some are proposing, how can a utility protect itself against collusion among bidders? What will be a utility's obligation if not enough capacity is bid? What would be the price for power from any capacity it is required to build? Will a power purchase strategy better shield the utility from ex post facto prudency and "used and useful" disallowances than a decision to build its own generating capacity?

Given the capital-intensive character of power production facilities, investors will undoubtedly insist on assurance of revenue sufficient to provide for a return on the investment made. About the only way IPPs can provide this assurance is with some form of collateral, such as a take-or-pay contract with a franchised utility. What will be the effect on the utility's financing flexibility if such long-term take-or-pay capacity commitments are regarded as debt equivalent obligations on its balance sheet?

TRANSMISSION ACCESS

The key issue for those who are expounding the need for increased competition in the electric utility industry is access to the transmission grid, especially "wheeling," or the third-party transfer of electricity between buying and selling utilities. Increased economic efficiency or lowering of aggregate electricity costs through wheeling is a commendable goal, as is reducing regional price differentials through increased access to low-cost interregional power suppliers. But what about the merits of transactions that merely transfer benefits from one customer class to another? Suppose, for example, wholesale municipal distributors or large industrial consumers were allowed to bypass their local utility service and purchase cheaper power. Would they be required to pay a wheeling charge that reflects the host utility's continuing and unavoidable obligation to serve and, therefore, will the utility be financially indifferent to the transaction? Or will captive customers end up covering the cost of stranded investments through higher tariffs?

Large regional price differentials and variations in generating reserve margins, along with the difficulty in obtaining new transmission rights-of-way, are likely to increase interregional power flows and force more intensive use of existing transmission corridors, according to the North American Electric Reliability Council. A recently released statement by the Institute of Electrical and Electronics Engineers (IEEE) describes the more significant technical considerations related to deregulation of the transmission grid (Tackaberry, 1988). The IEEE statement notes that the economic results from any restructuring of the electric power industry will depend on how well the essential technical considerations are accommodated. For this reason the IEEE statement strongly recommends that full consideration be given by public policymakers to technical, reliability, and safety factors—as well as theoretical economic factors—when restructuring proposals are being evaluated. The statement underscores the important economic benefits currently being realized through coordinated planning and operation of the highly integrated electric generation and transmission systems. These benefits result from reduced generation capacity margin requirements and the ability to schedule generation on a lowest incremental cost basis over broad geographical areas on a regional or multiregional power pool. The statement identifies operational stability of interconnected systems, fault (short-circuit) protection, power flow, generation scheduling, dispatch and control generation maintenance, voltage and frequency regulation, reactive power requirements, backup capacity, and emergency operating procedures as among the many technical factors that require carefully coordinated planning and operation of electric utilities on a regional or multiregional basis. The statement calls attention to the fact that the flow of electrical energy through a transmission network is determined by the electrical characteristics of the network at any moment in time; therefore, interchange or wheeling transactions may significantly affect the owners of network segments who are not parties to transactions between the buyer and the seller.

According to the North American Electric Reliability Council (1987), present and planned transmission system capability is heavily committed to capacity and energy transactions among utilities. To accommodate transactions beyond those already planned—with either utilities or nonutility generators—utilities need to build additional transmission facilities. But how will these be financed? Will regulators authorize charges for long-term wheeling service based on long-run marginal costs, or will third-party wheeling customers be given access to the remaining network capacity at low embedded-cost-based tariffs? The latter arrangement may cause the retail customers of the utilities to bear the cost of new facilities. If new capacity cannot be built, how will existing capacity be allocated? How will utilities that are not a party to power contracts be compensated for inadvertent power flows over the transmission network? How will the liability for a blackout caused by overload or instability because of third-party access be allocated among the parties? These questions are more than technical, or even institutional; they are societal.

ROLE OF TECHNOLOGY

Another major driving force in shaping the electric power industry is the application of new technology on the supply side of the utility meter. New technology promises to increase the flexibility and efficiency of electric power production and delivery systems, and thus provide added resilience to accommodate a more competitive bulk power market with a larger number of unregulated buyers and sellers.

The most important development over the past two decades has been the great leap forward in information technology brought about by the convergence of rapid advances in microelectronics, computers, and telecommunications. These technologies are being widely applied to the design and operation of power systems. Solid-state devices are replacing mechanical relays for fault protection, and computers and microprocessors have taken over many data logging and control functions. Computer simulation is used widely for system planning, engineering, and operator training. In the future, artificial intelligence techniques may soon provide expert guidance for system operators, and robots will be used increasingly for maintenance tasks in hostile environments.

The United States is headed toward power systems that are electronically, rather than mechanically, controlled and thus are far more flexible than current utility networks. Most of the progress in electronics during the past several decades has involved use of low-voltage, low-current solid-state circuits, or chips. Power electronics applies this more than 30-year-old science to high-voltage, high-current, "power" applications by using solid-state devices called thyristors. These devices are capable of handling thousands of volts and hundreds of amperes.

Thyristors permit the switching of power circuits at the speed of light and thus offer a means of control that is vastly superior to mechanical switching. This technology should permit the United States to enhance use of the transmission grid substantially without compromising system reliability (Hingorani, 1987). The thyristor and its variations are now being used in AC/DC converters for high-voltage transmission of direct current, static reactive power compensators, uninterruptible power supplies to protect sensitive equipment, and drives for adjustable-speed motors. Power semiconductor devices can ease the problem of feeding the direct current generated by photovoltaic facilities, fuel cells, and storage batteries into the alternating current power lines that are universally employed by utilities throughout the world. In the future, there is the prospect of an all-thyristor-based phase shifter and ultimately a device that can rapidly control both active and reactive power flows through utility networks. Thyristors may eventually replace mechanical circuit breakers on utility distribution power lines.

In the more distant future, high-transition-temperature superconductivity, high-performance fission energy, and solar power conversion systems offer promise for still further improvement in the cost and reliability of the U.S. electric power supplies.

CONCLUSION

Rationalization of the issues concerning U.S. electric power systems is central to the continued electrification scenario. Constructive political responses will require that regulators, government officials, opinion leaders, and the public understand the economic issues and technological constraints involved in deregulating the nation's electric power systems as well as the opportunities afforded by new technology.

REFERENCES

Federal Energy Regulatory Commission. 1988. Notice of Proposed Rule Making: RM 88-4 Regulations Governing Independent Power Producers, March 16, 1988; RM 88-5 Regulations Governing Bidding Programs, March 16, 1988; RM 88-6 Administrative Determination of Full Avoided Cost, Sale of Power to Qualifying Facilities and Interconnection Facilities. March 16. Washington, D.C.

Hingorani, N. 1987. Future opportunities for electric power systems. IEEE Power Engineering Review 7(October):4–5.

North American Electric Reliability Council. 1987. Reliability Assessment: The Future of Bulk Electric System Reliability in North America 1987–1996. Princeton, N.J.: North American Reliability Council.

Tackaberry, W. R. 1988. The role of electric power engineers in restructuring the electric power systems in the United States. IEEE Power Engineering Review 8(April):11.

Energy: Production, Consumption,
and Consequences. 1990.
Pp. 246–264. Washington, D.C.:
National Academy Press.

Regional Approaches to Transboundary Air Pollution

PETER H. SAND

International air pollution problems have traditionally been attributed to emissions from "point sources" being transported across national borders and have been viewed mainly as a matter for bilateral arrangements between neighboring countries. The bilateral approach is well illustrated by the U.S.-Canadian *Trail Smelter* arbitration (Read, 1963; United Nations, 1949), by the recent U.S.-Mexican agreement on transboundary air pollution caused by copper smelters in the common border area (Applegate and Bath, 1986; Utton, 1987), and by a number of similar local agreements or judicial settlements in Europe, such as the French-German case of *Poro* v. *Lorraine Basin Coalmines* (Bunge, 1986; Sand, 1974). Accordingly, the customary principle of "good neighborliness" (Goldie, 1972) was long considered as an adequate legal basis, both for intergovernmental arrangements and for granting private remedies to individual pollution victims across national boundaries (McCaffrey, 1975; Sand, 1977).

The advent of long-range transboundary air pollution (LRTAP) quickly shattered this confidence. First alerted by Scandinavian reports of "acid rain" damage due to air pollution from Western and Central Europe (Russell and Landsberg, 1971), the Cooperative Program for Monitoring and Evaluation of the Long-Range Transmission of Air Pollutants in Europe (EMEP) was established in 1977 (United Nations Economic Commission for Europe [UN/ECE], 1982). It has produced voluminous and increasingly reliable evidence that sulfur and nitrogen compounds emitted by a wide range of stationary and mobile pollution sources are dispersed through the atmosphere over thousands of miles. Because these sources are principally related to fossil fuel combustion, they are inextricably linked to energy consumption.

The EMEP program has three main elements (Dovland, 1987): (1) collection of emission data; (2) measurement of air and precipitation quality; and (3) modeling of atmospheric dispersion, using emission data, meteorological data, and functions describing the transformation and removal processes. The purpose of the models is to provide concentration and deposition profiles for major air pollutants over Europe. Coordination and intercalibration of chemical measurements are carried out at the Norwegian Air Research Institute in Lillestrøm; the two coordinating centers for modeling activities are the Norwegian Meteorological Institute in Oslo and the Institute for Applied Geophysics in Moscow. The EMEP sampling network, consisting of 95 stations in 24 countries of Western and Eastern Europe, is based on 24-hour sampling of air and precipitation. The accuracy of the atmospheric dispersion calculations is evaluated by frequent comparison with measurements. Canada and the United States also contribute to the program with reports and interlaboratory comparisons.

Figures 1 and 2 show the geographical distribution of sulfur dioxide (SO_2) and nitrogen dioxide (NO_2) emissions in Europe (including the European part of the USSR) on the standard sample grid, based on official emission data reported by governments. Figures 3 and 4 display the mean annual concentration of sulfate ($SO_4^=$, corrected for sea salt) and nitrate (NO_3^-), which are the main ions contributing to precipitation acidity, shown in Figure 5. Figure 6 shows the concentration of ozone (O_3), another important atmospheric pollutant caused by nitrogen oxides and hydrocarbon emissions in the region.

EMEP results—annually reviewed, updated, and approved by an intergovernmental steering body—make it possible to quantify the pollutant depositions in each country that can be attributed to emissions in any other country. Study of these matrices shows that for about half of the European countries concerned, the major part of the total pollutant deposition originates from foreign emissions (Eliassen et al., 1988). Even minor emitters such as Switzerland "export" some of their airborne pollution to neighboring countries such as Italy and as far afield as the Soviet Union. However, less than 15 percent (i.e., 17,000 metric tons) of the estimated 121,000 metric tons of sulfur deposited in Switzerland in 1980 originated from Swiss sources: 65,000 metric tons were "imported" by air from neighboring countries (Italy 31,000, France 23,000, Federal Republic of Germany 10,000, Austria 1,000). Another 20,000 metric tons originated in countries as distant as the United Kingdom, the German Democratic Republic, and Spain (4,000 each), Belgium, Czechoslovakia, and Poland (2,000 each), and Hungary and the Netherlands (1,000 each); the remainder came from indeterminate sources.

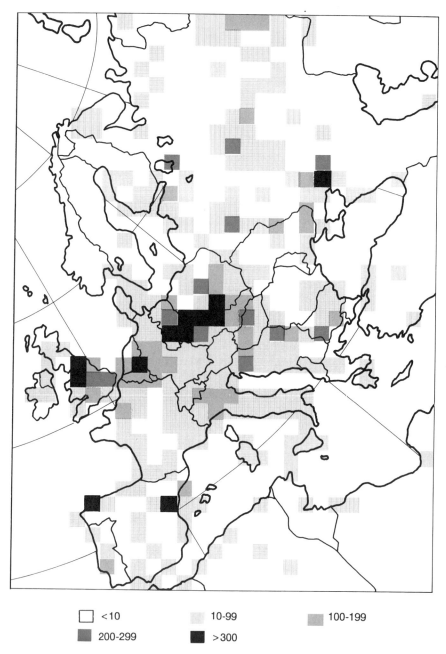

FIGURE 1 Emissions of sulfur dioxide in Europe, 1985 (in 1,000 metric tons of sulfur per year).

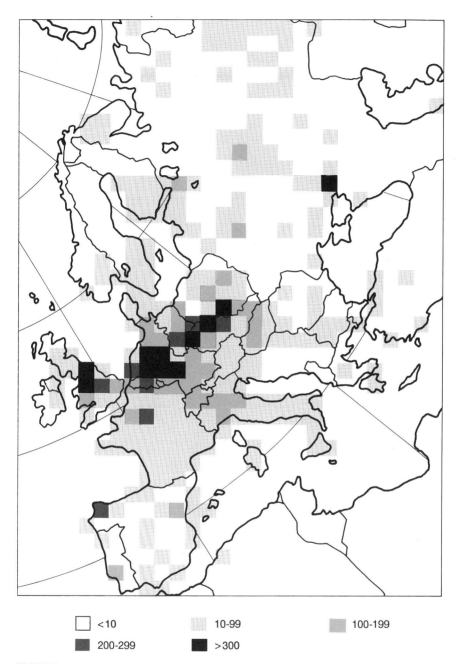

FIGURE 2 Emissions of nitrogen oxides in Europe, 1985 (in 1,000 metric tons of NO_2 per year).

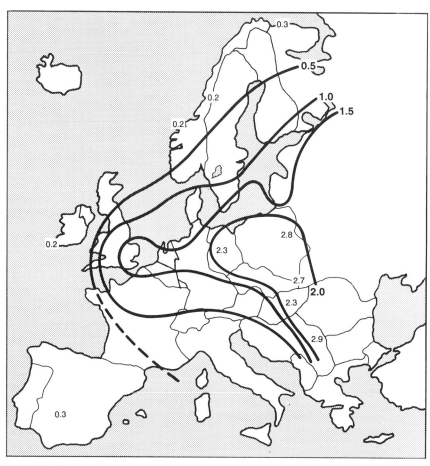

FIGURE 3 Mean annual concentration of sulfate in precipitation in Europe, 1985 (milligrams S per liter).

These findings demonstrated that bilateral concepts of geographical contiguity and "good neighborliness"—useful as they may have been for point-source transboundary pollution—had ceased to be sufficient or adequate for the region, and that a new multilateral approach was needed. Although international efforts to cope with long-range air pollution were at first confined to Western Europe—primarily the Organization for Economic Cooperation and Development in Paris (Eliassen, 1978), but also the Council of Europe in Strasbourg and the European Economic Community in Brussels (Adinolfi, 1968; Ercman, 1986; Smeets, 1982)—it soon became ev-

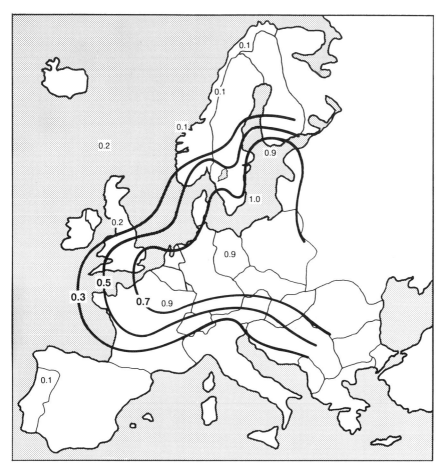

FIGURE 4 Mean annual concentration of nitrate in precipitation in Europe, 1985 (milligrams N per liter).

ident that the problem called for joint action on a wider scale and required cooperation between Western and Eastern European countries (Amendola and Sand, 1975; Lykke, 1977). The only intergovernmental body at this level was the UN Economic Commission for Europe in Geneva, one of the five regional commissions of the United Nations (Szasz and Willisch, 1983). It includes not only all European countries but also the United States and Canada and had already begun to deal with environmental problems in 1968 (Bishop and Munro, 1972; Stein, 1972).

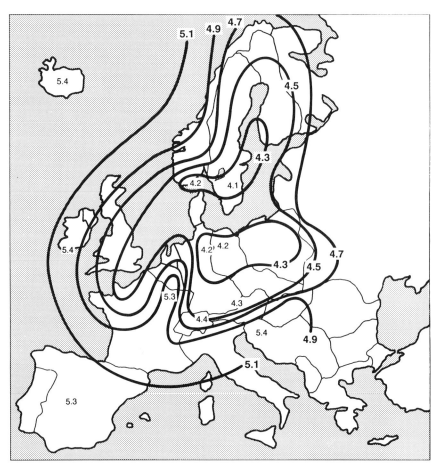

FIGURE 5 Mean annual precipitation acidity in Europe, 1985 (pH values).

JOINT PROGRAMS AND AGREEMENTS

After intensive diplomatic preparations and negotiations (Chossudovsky, 1988; Wetstone and Rosencranz, 1983), a Convention on Long-Range Transboundary Air Pollution was adopted in Geneva on November 13, 1979 (LRTAP Convention, 1979). It went into force on March 16, 1983, and currently has a membership of 32 parties including the United States and Canada (Frankel, 1989; Rosencranz, 1981; Sion, 1981). In addition to laying down common policy principles—including a commitment to use "the best available technology that is economically feasible" for air pollution abatement—this regional treaty establishes institutions for permanent cooperation, including annual review meetings of the intergovernmental

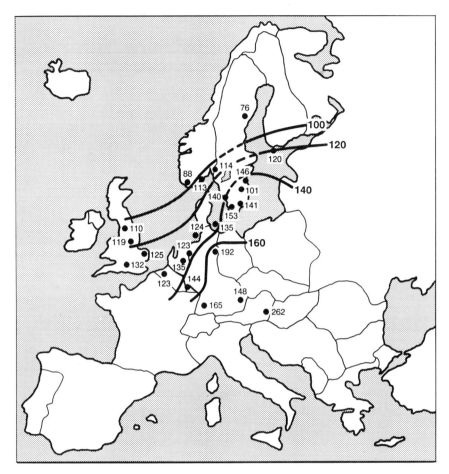

FIGURE 6 Ninety-eighth percentile of ozone concentrations in Europe from April to September 1985 (in micrograms per cubic meter).

Executive Body. Moreover, it provides the legal framework for several joint programs and specific agreements (Sand, 1987; Tollan, 1985):

1. A comprehensive review of national strategies and policies for air pollution abatement is undertaken every four years to ascertain the extent to which the objectives of the convention have been met. The first of these major "audits" of compliance was carried out in 1986 (UN/ECE, 1987a); it provides country-by-country tables of the most recent emission statistics, data on applicable national standards, and an overview of regulatory and economic measures taken by each party to comply with the treaty.

2. The EMEP Protocol was signed in September 1984, ratified by 31 parties (including the United States and Canada), and has been in force since January 28, 1988 (EMEP Protocol, 1984). It ensures the joint long-term financing of the above-mentioned regional monitoring program and its three international centers, with an annual budget of slightly more than $1 million, to which the United States and Canada currently contribute $10,000 each on a voluntary basis. Programs and budgets are supervised by the EMEP Steering Body and are approved at the annual meetings of the Executive Body for the Convention.

3. In addition, optional cooperative programs have been set up by the parties to the convention to monitor and assess air pollution effects on four main targets: *forests* (coordinated by the Federal Republic of Germany and by Czechoslovakia, with the first large-scale surveys of forest decline carried out in 25 European countries in 1986–1988); *materials and monuments* (coordinated by Sweden, with a uniform four-year exposure experiment using 39 test sites in 14 countries, including the United States, started in 1987); *rivers and lakes* (coordinated by Norway, with a first geographical survey of acidified surface waters in 21 countries, including Canada and the United States, carried out in 1988); and *agricultural crops* (coordinated by the United Kingdom, with a uniform controlled exposure experiment in 10 countries, including the United States, started in 1988). A new pilot program on integrated (i.e., multimedia and cross-media) monitoring, led by Sweden, also became operational in 1988. After endorsement by the Executive Body, scientific results of these programs are published regularly in the trilingual *Air Pollution Studies* series (UN/ECE, 1984–1989).

4. The Sulfur Protocol was adopted in Helsinki, Finland, on July 8, 1985 (Sulfur Protocol, 1985). It was ratified by 19 parties (including Canada but not the United States) and has been in force since September 2, 1987. It commits governments to reduce national emissions of sulfur compounds or their transboundary fluxes by at least 30 percent by 1993 at the latest, using 1980 as the baseline year (Björkbom, 1988; Vygen, 1985). As of 1988, 12 parties had reached the 30 percent target ahead of schedule; 10 parties are committed to cutting their SO_2 emissions below one-half of their 1980 levels by 1995, and at least four of these (Austria, the Federal Republic of Germany, Liechtenstein, and Sweden) to below one-third. For large combustion plants in particular, the environmental ministers of the European Economic Community (EEC)—after reaching a compromise in Brussels on June 17, 1988, that grants exemptions to certain member countries such as Ireland, Spain, and the United Kingdom—agreed to accelerate the schedule for reducing sulfur emissions (European Economic Community, 1988). The agreement calls for reductions of 40 percent by 1993 (UK 20 percent), 60 percent by 1998 (UK 40 percent), and 70 percent by 2003 (UK 60 percent). Earlier EEC decisions and

directives had already tightened the standards for sulfur content in fuel (Oil Companies' European Organization for Environmental and Health Protection [CONCAWE], 1989).

5. A new Protocol on Nitrogen Oxides, signed by 25 countries, including the United States and Canada, at the sixth session of the Executive Body in Sofia, Bulgaria, on November 1, 1988, calls for a "freeze" of national NO_x emissions or their transboundary fluxes at 1987 levels by 1994, to be followed by reductions from 1996 onward at a rate yet to be agreed (Nitrogen Oxides Protocol, 1988). The United States exercised an option to select an earlier baseline year (namely, 1978), with the understanding that this will not result in an increase of emission averages or transboundary fluxes over 1987 levels by 1996. On the other hand, the Sofia Protocol expressly reserves the right of parties to take stricter regulatory measures individually or collectively. Twelve European signatories thus went a step further and signed an additional (and separate) joint declaration committing them to a 30 percent reduction of emissions by 1998. The protocol-plus-declaration formula helped to resolve a dilemma typical of multilateral treaty-making in the environmental field: ensuring (a) to get as many countries as possible to join when wide geographical coverage is essential, without (b) slowing down the convoy to the speed of the slowest boat.

6. Next in line for international regulation will be emissions of volatile organic compounds (VOCs). Considered one of the main causes of pollution-related forest decline in Europe, VOCs, together with NO_x, are precursors in the photochemical formation of tropospheric ozone and other pollutant oxidants (Nilsson and Duinker, 1987; Prinz, 1987). The Sofia meeting therefore set up a new VOC Working Group (chaired by France) to evaluate the scientific evidence and to prepare proposals for a further protocol.

Taken together, these agreements and programs may be described as an international regime, namely, "principles, norms, rules, and decision-making procedures around which actor expectations converge in a given issue-area" (Haas, 1980; Krasner, 1983). How effective has this "functional regionalism" (Majone, 1986) been in terms of actual pollution trends?

As shown in Figure 7—using official data and forecasts reported by all parties to the Convention, whether or not they ratified the Sulfur Protocol—there has definitely been an overall net decline in sulfur dioxide emissions since 1980. Moreover, this trend is expected to continue in the 1990s despite rising energy demands. Figure 8 is a first attempt at correlating these recent figures with long-term assessments of energy consumption and SO_2 emission trends in Europe (Dovland and Semb, 1980; Eliassen et al., 1988; Fjeld, 1976).

It may be too early to interpret these figures as evidence of actual "decoupling" of energy consumption from energy pollution trends. It should also be kept in mind that at least part of the emission reductions can be attributed to fuel switches from coal and oil to increased use of nuclear energy and natural gas for power production, for example, in France and several East European countries. Nevertheless, the rising impact of investments and improvements in emission control technology can also be documented, for example, in the review of national implementation of the Convention up to 1986 (UN/ECE, 1987a). Regular information exchange within the framework of the Convention—for example, the 1986 Graz Seminar on the Control of Sulfur and Nitrogen Oxides from Stationary Sources (UN/ECE, 1987b)—has not only promoted the dissemination of new antipollution technologies but also tends to strengthen consensus on their technical and economic "feasibility" (LRTAP Convention, article 6). A key factor in this process of mutual education has been the active participation of nongovernmental organizations, including both industrial groups (e.g., the International Chamber of Commerce, the International Union of Producers and Distributors of Electric Energy, the International Road

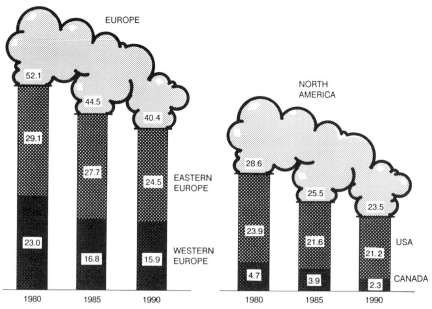

FIGURE 7 Trends in total man-made emissions of sulfur dioxide in Europe and North America, 1980–1990 (million metric tons of SO_2 per year).

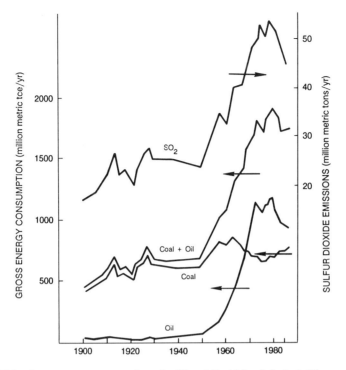

FIGURE 8 Gross energy consumption of solid and liquid fossil fuels (million metric tons of coal equivalent) and total sulfur dioxide emissions (million metric tons of SO_2 per year) in Europe, 1900–1985.

Transport Union, and CONCAWE) and environmental groups (e.g., Greenpeace, Friends of the Earth, and the International Union for Conservation of Nature and Natural Resources).

POLICY CRITERIA FOR INTERNATIONAL AGREEMENTS

Paradoxically, perhaps, the very open-endedness and generality of the LRTAP Convention's provisions—frequently pointed out by its critics (Gündling, 1986; McCormick, 1985)—may be one of its main assets, as new priorities for action continue to arise and are met in turn by the flexible "protocol" system. Even though the convention binds all parties to common goals and institutions, participation in protocols is optional. As with any international agreement, it is up to each party to weigh the pros and cons of joining or staying outside. Besides environmental criteria, governments usually evaluate agreements in the light of many other factors, including (a) economic costs and benefits; (b) compliance controls and verification;

and (c) equity for all parties concerned. These policy criteria can also be applied to the sulfur and nitrogen protocols under the LRTAP Convention.

Economic Rationale

Any environmental benefits of a treaty regime, however quantified, invariably mean pollution control costs—costs to governments and costs to industry. In an international setting, countries will first seek to avoid penalizing their own domestic industries and taxpayers; rather than imposing such costs unilaterally, they will instead endeavor to ensure that costs are spread evenly and, in particular, are also imposed on foreign competitors. Hence, one way of looking at intergovernmental environmental agreements is as legal mechanisms to ensure a degree of international equalization of costs among competing polluters. In this perspective—with Japan already leading North America in regulatory requirements for control of sulfur and nitrogen emissions (Weidner, 1986)—the focus of attention is on Europe. Whereas Austria, Switzerland, and the Nordic countries already require that new emission controls for motor vehicles match 1987 U.S. federal standards (Assarsson, 1986; Walsh, 1988), specifications applicable in the 12 member states of the European Economic Community are roughly comparable to U.S. standards of 1983 (CONCAWE, 1989; Lomas, 1988), gradually to be tightened in 1991 and 1993. The Nitrogen Oxides Protocol accelerates this process by "technology-forcing" provisions requiring use of the best available and economically feasible technology, as specified in the protocol's technical annex.

One nonnegligible side effect is gradual international standardization in the growing market for pollution control technology. Uniform emission standards for motor vehicles are set and periodically revised under a 1958 agreement administered by the United Nations Economic Commission for Europe (UN/ECE, 1958). In actual practice, many of these standards are followed not only by the 22 West and East European countries having signed the agreement but also by a growing number of other countries, as a convenient shortcut to national type-approval of imported foreign cars and spare parts. The Nitrogen Oxides Protocol, besides using these standards as a baseline (in its technical annex on mobile emission sources) also opens the way for wider application of new technologies such as catalytic converters, by requiring all parties to make unleaded fuel available "as a minimum along main international transit routes" as soon as possible and no later than two years after the date on which the protocol enters into force.

Verification

Another major concern of governments contemplating accession to an international treaty is usually compliance control: regular monitoring to ensure that each party lives up to its commitments. In this respect, the LRTAP Convention probably is unique, with periodic reviews of state practice carried out every four years, and with the EMEP monitoring network, which has now been in operation for more than 10 years, internationally accepted as a reference source by participating governments and by the scientific community (Eliassen and Saltbones, 1983; Gosovic, 1989; Israel et al., 1987; United Nations Environment Program [UNEP], 1986; Wallén, 1986). Few other international agreements can be said to come equipped with verification instruments of this caliber.

Weighing the Equities

Unlike bilateral transboundary agreements—where asymmetries and even inequities are often resolved by ad hoc trade-offs completely unrelated to the alleged environmental purposes of a treaty (Carroll, 1983; Handl, 1986)—a stable regional regime must be built on strict reciprocity and fairness to all parties concerned. The needed balance can be achieved by several alternative means, as illustrated by the LRTAP protocols on sulfur and on oxides of nitrogen.

The Sulfur Protocol uses a flat-rate percentage reduction comparable to the "bubble" approach of domestic air quality regulation in the United States and the Federal Republic of Germany. This approach was also used in the more recent Montreal protocol on ozone-depleting substances, concluded in September 1987 under the auspices of the UNEP (Széll, 1988). All emissions from a country are aggregated annually; as long as the overall reduction can be met by the prescribed deadlines, each country is free to choose its own methods for reaching that target. These methods may include cutting emissions from all sources proportionately, or imposing heavier cuts selectively on major emission sources or source categories and then allowing for internal deals and "bubble trades" between them. The percentage is uniform for all parties (30 percent of 1980 levels by 1993), but countries may opt to measure and reduce their annual SO_2 transboundary fluxes rather than total national emissions.

The Nitrogen Oxides Protocol adopts a uniform flat-rate "freeze" as a first step. It takes a more flexible approach at the second stage, where emission cuts are to be specified selectively on the basis of internationally accepted and periodically revised critical loads. Although international standard-setting for environmental purposes is a well-established practice (Contini and Sand, 1972; Sand, 1980), the "critical loads" concept is comparatively new in air pollution control (Ekman, 1986; Nilsson, 1986). The closest analogy is the "safe minimum standards" concept used in natural resources management (Ciriacy-Wantrup, 1968). Translating these standards into actual control options and measures will require a sequence of further international and national decision-making, also taking into account elements of welfare economics, including comparative cost-effectiveness and optimal resource allocation (Alcamo et al., 1987; Hordijk, 1988; Persson, 1988). Elaboration of the critical loads concept is part of the mandate given to a new Working Group on Abatement Strategies established for this purpose by the Executive Body for the Convention at its 1988 Sofia session.

Even as the regional regime of transboundary air pollution control under the Geneva Convention continues to evolve, the basic concern of governments remains essentially as formulated more than 80 years ago by the U.S. Supreme Court in the case of *State of Georgia* v. *Tennessee Copper Company and Ducktown Sulphur, Copper and Iron Company, Ltd.* (U.S. Supreme Court, 1907):

> It is a fair and reasonable demand on the part of a sovereign that the air over its territory should not be polluted on a great scale by sulphurous acidic gas; that the forests on its mountains, be they better or worse, and whatever domestic destruction they may have suffered, should not be further destroyed or threatened by the acts of persons beyond its control; and that the crops and orchards on its hills should not be endangered from the same source.

NOTE

Views and opinions expressed are those of the author and do not necessarily reflect those of UN/ECE.

REFERENCES

Adinolfi, G. 1968. First steps toward European cooperation in reducing air pollution: Activities of the Council of Europe. Law and Contemporary Problems 33:421-426.

Alcamo, J., M. Amann, J. P. Hettelingh, M. Holmberg, L. Hordijk, J. Kämäri, L. Kauppi, P. Kauppi, G. Kornai, and A. Mäkelä. 1987. Acidification in Europe: A simulation model for evaluating control strategies. Ambio 16:232-245.

Amendola, G., and P. H. Sand. 1975. Transnational environmental cooperation between different legal systems in Europe. Earth Law Journal 1:189-195.

Applegate, H. G., and C. R. Bath. 1986. Air pollution in a transboundary setting: The case of El Paso, Texas, and Ciudad Juarez, Chihuahua. Pp. 96-116 in Transboundary Air Pollution: International Legal Aspects of the Cooperation of States, C. Flinterman, B. Kwiatkowska, and J. G. Lammers, eds. Dordrecht, Netherlands, and Boston, Mass.: Martinus Nijhoff Publishers.

Assarsson, B. 1986. Swedish policy on exhausts. Acid Magazine 4:8-9. Stockholm: Swedish Environmental Protection Board.

Bishop, A. S., and R. D. Munro. 1972. The UN regional economic commissions and environmental problems. International Organization 26:348-371.

Björkbom, L. 1988. Resolution of environmental problems: The use of diplomacy. Pp. 123-137 in International Environmental Diplomacy, J. E. Carroll, ed. Cambridge, England: Cambridge University Press.

Bunge, T. 1986. Transboundary cooperation between France and the Federal Republic of Germany. Pp. 181-198 in Transboundary Air Pollution: International Legal Aspects of the Cooperation of States, C. Flinterman, B. Kwiatkowska, and J. G. Lammers, eds. Dordrecht, Netherlands, and Boston, Mass.: Martinus Nijhoff Publishers.

Carroll, J. E. 1983. Environmental Diplomacy: An Examination and a Prospective of Canadian-U.S. Transboundary Environmental Relations. Ann Arbor, Mich.: University of Michigan Press.
Chossudovsky, E. M. 1988. "East-West" Diplomacy for the Environment in the United Nations. E.88/XV/ST26. Geneva: United Nations Institute for Training and Research.
Ciriacy-Wantrup, S. V. 1968. Resource Conservation: Economics and Policies, 3rd ed. Berkeley: University of California Division of Agricultural Sciences.
Contini, P., and P. H. Sand. 1972. Methods to expedite environment protection: International ecostandards. American Journal of International Law 66:37–59.
Dovland, H. 1987. Monitoring European transboundary air pollution. Environment 29(10):10–15, 27–28.
Dovland, H., and A. Semb. 1980. Atmospheric transport of pollutants. Pp. 14–21 in Proceedings of the International Conference on the Ecological Impact of Acid Precipitation, D. Drablös and A. Tollan, eds. Oslo, Norway: SNSF Project.
Ekman, H. 1986. The limits to nature's tolerance. Acid Magazine 4:28–30. Stockholm: Swedish Environmental Protection Board.
Eliassen, A. 1978. The OECD study of long-range transport of air pollutants: Long-range transport modelling. Atmospheric Environment 12:479–487.
Eliassen, A., and J. Saltbones. 1983. Modelling of long-range transport of sulphur over Europe: A two-year model run and some model experiments. Atmospheric Environment 17:1457–1473.
Eliassen, A., O. Hov, T. Iverson, J. Saltbones, and D. Simpson. 1988. Estimates of airborne transboundary transport of sulphur and nitrogen over Europe. EMEP/MSC-W Report 1/88. Oslo, Norway: Norwegian Meteorological Institute.
EMEP Protocol. 1984. Protocol to the 1979 Convention on Long-Range Transboundary Air Pollution, on long-term financing of the Cooperative Program for Monitoring and Evaluation of the Long-Range Transmission of Air Pollutants in Europe (EMEP). United Nations Document ECE/EB.AIR/11. Reprinted in International Legal Materials 24(1985):484.
Ercman, S. 1986. Activities of the Council of Europe and the European Economic Communities related to transboundary air pollution. Pp. 131–140 in Transboundary Air Pollution: International Legal Aspects of the Cooperation of States, C. Flinterman, B. Kwiatkowska, and J. G. Lammers, eds. Dordrecht, Netherlands, and Boston, Mass.: Martinus Nijhoff Publishers.
European Economic Community (EEC). 1988. Council directive of 24 November 1988 on the limitation of emissions of certain pollutants into the air from large combustion plants (88/609/EEC). Official Journal of the European Communities (L 336):1.
Fjeld, B. 1976. Forbruk av fossilt brensel i Europa og utslipp av SO_2 i perioden 1900–1972 (Fossil fuel consumption in Europe and SO_2 emissions during the period 1900–1972). NILU report TN1/76. Lilleström, Norway: Norwegian Institute for Air Research.
Frankel, A. 1989. Convention on Long-Range Transboundary Air Pollution. Harvard International Law Journal 30:447–477.
Goldie, L. F. E. 1972. Development of an international environmental law: An appraisal. Pp. 104–165 in Law, Institutions, and the Global Environment, J. L. Hargrove, ed. Dobbs Ferry, N.Y.: Oceana Publications.
Gosovic, B. 1989. Earthwatch for the Twenty-First Century: Piecing Together a Global Environment Monitoring System. London: Rutledge.
Gündling, L. 1986. Multilateral cooperation of states under the ECE Convention on Long-Range Transboundary Air Pollution. Pp. 19–31 in Transboundary Air Pollution: International Legal Aspects of the Cooperation of States, C. Flinterman, B. Kwiatkowska, and J. G. Lammers, eds. Dordrecht, Netherlands, and Boston, Mass.: Martinus Nijhoff Publishers.
Haas, E. B. 1980. Why collaborate? Issue-linkage and international regimes. World Politics 32:357–405.

Handl, G. 1986. Transboundary resources in North America: Prospects for a comprehensive management regime. Pp. 63–93 in Transboundary Air Pollution: International Legal Aspects of the Cooperation of States, C. Flinterman, B. Kwiatkowska, and J. G. Lammers, eds. Dordrecht, Netherlands, and Boston, Mass.: Martinus Nijhoff Publishers.

Hordijk, L. 1988. Linking policy and science: A model approach to acid rain. Environment 30(2):17–20, 40–42.

Israel, Y. A., I. M. Nazarov, and S. D. Fridman, eds. 1987. Monitoring Transgranichnogo Perenosa Zagryaznyayushchikh Vozdukh Vyeshchestv (Monitoring the transboundary transmission of air pollutants). Leningrad, USSR: Hydro-Met Publications.

Krasner, S. D., ed. 1983. International Regimes. Ithaca, N.Y.: Cornell University Press.

Lomas, O. 1988. Environmental protection, economic conflict and the European Community. McGill Law Journal 33(3):506–539.

LRTAP Convention. 1979. Convention on Long-Range Transboundary Air Pollution. United Nations Document E/ECE/1010. United States Treaties and International Agreements Series No. 10541. Reprinted in International Legal Materials 18:1442.

Lykke, E. 1977. International effort needed to combat airborne pollution. European Free Trade Association (EFTA) Bulletin 18(7):15–17.

Majone, G. 1986. International institutions and the environment. Pp. 351–358 in Sustainable Development of the Biosphere, W. C. Clark and R. E. Munn, eds. Cambridge, England: Cambridge University Press.

McCaffrey, S. C. 1975. Private Remedies for Transfrontier Environmental Disturbances. IUCN Environmental Policy and Law paper no. 8. Morges, Switzerland: International Union for Conservation of Nature and Natural Resources.

McCormick, J. 1985. Acid Earth: The Global Threat of Acid Pollution. Earthscan, J. Tinker, ed. London and Washington, D.C.: International Institute for Environment and Development.

Nilsson, J., ed. 1986. Critical loads for nitrogen and sulphur. Miljö report no. 11. Stockholm, Sweden: Nordic Council of Ministers.

Nilsson, S., and P. Duinker. 1987. The extent of forest decline in Europe: A synthesis of survey results. Environment 29(9):4–9, 30–31.

Nitrogen Oxides Protocol. 1988. Protocol to the 1979 Convention on Long-Range Transboundary Air Pollution concerning the control of emissions of nitrogen oxides or their transboundary fluxes. United Nations Document ECE/EB.AIR/22. Reprinted in International Legal Materials 28(1989):214.

Oil Companies' European Organization for Environmental and Health Protection (CONCAWE). 1989. Trends in Motor Vehicle Emission and Fuel Consumption Regulations: 1989 Update. Report No. 6/89. The Hague, Netherlands: CONCAWE.

Persson, G. 1988. Toward resolution of the acid rain controversy. Pp. 189–196 in International Environmental Diplomacy, J. E. Carroll, ed. Cambridge, England: Cambridge University Press.

Prinz, B. 1987. Causes of forest damage in Europe: Major hypotheses and factors. Environment 29(9):11–15, 32–37.

Read, J. E. 1963. The Trail Smelter dispute. Canadian Yearbook of International Law 1:213–229.

Rosencranz, A. 1981. The ECE Convention of 1979 on Long-Range Transboundary Air Pollution. American Journal of International Law 75:975–982. Reprinted in Zeitschrift für Umweltpolitik 4:511–520.

Russell, C. S., and H. H. Landsberg. 1971. International environmental problems: A taxonomy. Science 172:1307–1314.

Sand, P. H. 1974. Transfrontier air pollution and international law. Pp. 107–113 in New Concepts in Air Pollution Research, J. O. Willums, ed. Experientia Supplementum No. 20. Basel, Switzerland: Birkhäuser Verlag.

Sand, P. H., 1977. The role of domestic procedures in transnational environmental disputes. Pp. 146–202 in Legal Aspects of Transfrontier Pollution, H. van Edig, ed. Paris: Organization for Economic Cooperation and Development.

Sand, P. H. 1980. The creation of transnational rules for environmental protection. Pp. 311–320 in Trends in Environmental Policy and Law, M. Bothe, ed. IUCN Environmental Policy and Law paper no. 15. Gland, Switzerland: International Union for Conservation of Nature and Natural Resources.

Sand, P. H. 1987. Air Pollution in Europe: International policy responses. Environment 29(10):16–20, 28–29, with a correction in Environment 30(1988)(2):42.

Sion, I. G. 1981. Regional approach to environmental protection and the UN/ECE Convention on Long-Range Transboundary Air Pollution. Revue Roumaine d'Etudes Internationales 15:317–400.

Smeets, J. 1982. Air quality limits and guide values for sulphur dioxide and suspended particulates—A European Community directive. Environmental Monitoring and Assessment 1:373–382.

Stein, R. E. 1972. The potential of regional organizations in managing man's environment. Pp. 253–293 in Law, Institutions, and the Global Environment, J. L. Hargrove, ed. Dobbs Ferry, N.Y.: Oceana Publications.

Sulfur Protocol. 1985. Protocol to the 1979 Convention on Long-Range Transboundary Air Pollution, concerning the reduction of sulfur emissions or their transboundary fluxes by at least 30 percent. United Nations Document ECE/EB.AIR/12. Reprinted in International Legal Materials 27(1988):707.

Szasz, P., and J. Willisch. 1983. Regional commissions of the United Nations. Encyclopedia of Public International Law 6:296–301. Amsterdam, Netherlands: North-Holland Publishing.

Széll, P. 1988. The Montreal Protocol on Substances that Deplete the Ozone Layer. International Digest of Health Legislation 39:278–282.

Tollan, A. 1985. The Convention on Long-Range Transboundary Air Pollution. Journal of World Trade Law 19:615–621.

United Nations (UN). 1949. The Trail Smelter Arbitration. Reports of International Arbitral Awards 3:1905–1982. United Nations sales no. 1949.V.2. New York.

United Nations Economic Commission for Europe (UN/ECE). 1958. Agreement concerning the adoption of uniform conditions of approval and reciprocal recognition of approval for motor vehicles equipment and parts. United Nations Treaty Series 335:211, 740:364 (as amended/supplemented through 1989).

UN/ECE. 1982. EMEP: The Cooperative Programme for Monitoring and Evaluation of the Long-Range Transmission of Air Pollutants in Europe. Economic Bulletin for Europe 34(1):29–40. United Nations: Pergamon Press.

UN/ECE. 1984–1989. Air Pollution Studies, nos. 1–5. United Nations sales nos. E.84.II.E.8, E.85.II.E.17, E.86.II.E.23, E.87.II.E.36, E.89.II.E.25. New York.

UN/ECE. 1987a. National strategies and policies for air pollution abatement: Results of the 1986 major review prepared within the framework of the Convention on Long-Range Transboundary Air Pollution. United Nations sales no. E.87.II.E.29. New York.

UN/ECE. 1987b. Technologies for control of air pollution from stationary sources. Economic Bulletin for Europe 39(1):1–244. United Nations: Pergamon Press.

United Nations Environment Program (UNEP). 1987. United Nations Environment Program: Environmental Data Report. Prepared for UNEP by the Monitoring and Assessment Research Centre, London. Oxford and New York: Basil Blackwell.

U.S. Supreme Court. 1907. State of Georgia versus Tennessee Copper Company and Ducktown Sulphur, Copper and Iron Company, Ltd. Supreme Court Reporter 206:230; 237:474, at page 477.

Utton, A. E., ed. 1987. Agreement of cooperation between the United Mexican States and the United States of America regarding transboundary air pollution caused by copper smelters along their common border, of January 29, 1987. Transboundary Resources Report 1(3):5–7.

Vygen, H. 1985. Air pollution control: Success of East-West cooperation. Environmental Policy and Law 15:6–8.

Wallén, C. C. 1986. Sulphur and nitrogen in precipitation: An attempt to use BAPMoN and other data to show regional and global distribution. WMO/TD-no. 103, Environmental Pollution Monitoring and Research Programme no. 26. Geneva, Switzerland: World Meteorological Organization.

Walsh, M. P. 1988. Worldwide developments in motor vehicles pollution control reflect persisting problems, varying standards, technological growth. International Environment Reporter 11:41–49.

Weidner, H. 1986. Japan: The success and limitations of technocratic environmental policy. Policy and Politics 14:43–70.

Wetstone, G. S., and A. Rosencranz. 1983. Acid Rain in Europe and North America: National Responses to an International Problem. Washington, D.C.: Environmental Law Institute.

Efficiency, Machiavelli, and Buddah

ROBERT MALPAS

A few years ago, in Japan, I came across the works of Buddha and found the following:

> There are two kinds of worldly passions that defile and cover over the purity of Buddha-nature. The first is the passion for analysis and discussion by which people become confused in judgment. . . . The second is the passion for emotional experience by which people's values become confused.

If ever there was a subject in danger of too much analysis and too much emotion at the expense of objectivity, that subject is energy. Of course we need analysis to help us understand, and we need well-directed emotion—even passionate emotion—to get things done. But we must not lose sight of the essential reality in the process: energy is fundamental to our material well-being.

Those in the richer countries want more of what energy makes possible to enable more travel; to heat, cool, and illuminate more buildings; to communicate more. The hundreds of millions of people in the poorer countries need more of what energy can provide just to improve the standard of their everyday life. Notice that energy is a means to many ends, and that it is the ends that are desired, not necessarily the energy itself. Yet most analysis, and certainly all of the passionate emotion, is directed to the supply of energy and very little to how more of its fruits can be obtained for less.

Associated with this clear need are mounting fears. The events of 1973 and 1979 aroused them worldwide—fears concerning the finiteness of oil and, hence, of high prices. The events of Three Mile Island and

Chernobyl raised fears that are severely restricting the growth of nuclear power. Lately, the fear of the greenhouse effect from burning fossil fuels has been spreading fast.

That is the overall picture. We need more of the fruits of energy, yet we increasingly fear the environmental effects from using it; we fear for its availability and sometimes its price. There are only a few global forces that affect the energy scene, but they are powerful. In contrast, there are many *national* forces that impinge mainly on the supply of energy.

From all this, a simple common theme suggests itself: the need for energy efficiency. This theme, with some notable exceptions, has been largely overlooked in a large number of supply-side energy analyses.

There is a clear option available that will help both to meet the needs and to reduce the fears that have been noted. It is to accelerate the rate at which energy efficiency is increasing, both in its supply and in its use. The oil shocks of 1973 and 1979 provided a great spur through the sudden price increases, but much more can be and needs to be done.

The trouble is that, at present, most global forces tend to retard the rate of advance. The purpose of this discussion is to urge the engineering profession to make the drive for greater energy efficiency a powerful global force. I speak as an engineer who works for an oil company. In pursuing a theme that encourages the world to use less of our products, you might wonder how I keep my job. Rest assured that even the most optimistic predictions on energy efficiency will offer more than enough business opportunities, not only to provide the necessary fuels but also to serve the needs that will arise from the quest for greater efficiency.

To produce more of what you want while using less of what you have is basic to the teaching of all engineers. It is part of our fundamental philosophy. So I would expect all engineers to embrace enthusiastically the policy options that emerge from this attempt to stimulate greater energy efficiency.

In the richer countries of the world, each person consumes the energy equivalent of 36 barrels of oil per year. In the United States, energy consumption is 55 barrels per person per year! One might think that policies that result in requiring only, say, 18 barrels to produce more of what is currently wanted by the year 2020 would be vigorously supported.

In the poorer countries, 6 barrels per person per year offer only a bare subsistence. If by the year 2020, that same 6 barrels could provide a significantly higher standard of living, one might think that this also would be a goal all would strive to achieve.

This will not happen, however, unless much more effort and attention are devoted to demand-side issues and policies. Consider the principal global forces affecting the energy scene. These include supply, in terms of both the quantities of energy available and where the resources are to be found, and demand, which determines how much is required. Technology is a major force that constantly shifts the goalposts of both supply and demand.

There are political forces that are almost entirely national but have international consequences. Concern about the environment and the world ecosystem is now a major public issue.

Then there are those two human factors—fear and basic need—that are powerful stimulators of all global forces. The fears have already been mentioned. Basic need requires no elaboration, except to say that it leads to decisions being taken on the basis of short-term considerations without any thought of long-term consequences. Unfortunately, the major energy issues are mostly long-term.

To complete the list of global forces affecting the energy scene, there is the reality of economics. On the macroscale, this encompasses the desire for economic growth vigorously sought after by all people, rich and poor. On the microscale, energy economics includes investment and revenue decisions at national, local, corporate, and individual levels. Let us review these forces briefly.

On the supply side, while accounting for less than half of the total energy demand, crude oil still dominates the energy scene. Some would say that its availability and price are the *only* global forces of any real significance. Certainly, the oil shocks of 1973 and 1979 bear out that claim. They aroused all the fears and hence were an enormous spur to energy efficiency.

Equally important, this greater energy efficiency uncoupled economic and energy growth. The one-to-one relationship that had lasted so long seemed like an incontrovertible law of economics. Breaking the link was a major event and contributed to the dramatic fall in the oil price in 1986.

Crude oil supplies are, however, finite, and although countries in the Organization for Economic Cooperation and Development (OECD) may at present have reduced their dependence on supplies from the Middle East, two-thirds of the world's oil reserves are still located there (Figure 1). The problem is that the drop in oil prices makes it difficult for the public to recognize this fact. Dire warnings are heeded only when manifested in high prices, not otherwise. Of course, low prices do not encourage investment for greater energy efficiency.

On the demand side, the realization that economic growth is no longer directly limited by energy supply has provided an opportunity, which is excellently articulated in the publication *Energy for a Sustainable World* (Goldemberg et al., 1987). This opportunity is undoubtedly a force in favor of greater efficiency, but it is still a small force, which must be strengthened.

Technology, of course, exerts a major force. The advent of microcircuits and the microprocessor has resulted in spectacular advances in

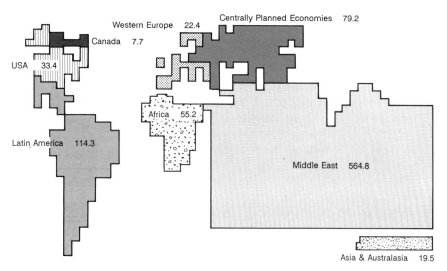

FIGURE 1 Proven oil reserves as of 1987 (billions of barrels).

technology, which will continue on their own momentum (e.g., advances in miles per gallon driven or flown, lumens per watt, degrees of heat in our buildings, units of information per unit of energy). Yet, paradoxically the very success of technology, because it reduces energy consumption, also reduces the economic incentive to invest further in greater efficiency.

On the supply side, technology has two major effects. It is reducing the price at which it is economic to discover and develop oil fields previously considered uneconomic, and it is reducing the price at which alternatives to crude oil become economic, such as harnessing very heavy oil or converting natural gas to gasoline. World resources of very heavy crude oil and natural gas are each as large as reserves of conventional crude oil.

The result of all this is best illustrated by long-term forecasts of the price of crude oil. Planners have become wiser: they now forecast a bracket. In 1986 the upper bound was $30 and the lower, $15. The rationale was that above $30, alternatives to conventional crude would become economical, and that below $15, demand would quickly revive and the economic penalties on suppliers would be too hard to bear. Today the upper bound has already been reduced to $25—in lower valued dollars at that.

Of course, we applaud these achievements and call for more; but they do undermine efforts toward higher energy efficiency. The public concludes that technology will come to its rescue on every issue. People believe that technology will continue to extend the finiteness of oil, as indeed it has, and that it will reduce the energy needed per unit of output, without any action or investment on their part. They also believe that environmental and ecological concerns will be solved by the cavalry—technology—riding over the hill! (Superconductivity seems to be the name of one of its younger officers!)

Lest anyone derives too much comfort from all this, it should be emphasized that even in the most optimistic energy-efficient scenario available, the United States will, by the year 2000, be importing more than half of its crude oil requirements.

Under the heading of technology one must recognize the remarkable increase in the use of electricity in the world. It is unquestionably the most convenient form of energy. It is intense—much more so than fossil fuels—and very easy to control, measure, and program. Its growth strongly favors greater energy efficiency.

On an international plane, politics is concerned with ensuring that the world is not unduly dependent on its supply of energy from a particular group of governments—whoever they may be. At the national level, politics is concerned with self-sufficiency, if possible, or less dependency if not. It means ensuring sufficient energy to sustain national growth, supplying the basic needs of the poor, and raising revenue by taxing energy. It also means protecting local environments and worrying about emissions from neighboring countries. It is predominantly concerned with supply issues. The following are some random examples.

Brazil, South Africa, and New Zealand have invested in expensive options to seek greater self-sufficiency: Brazil, in ethanol; South Africa in converting coal to gasoline; and New Zealand, in natural gas conversion to gasoline. These policies are now heavily subsidized because they were predicated on the expectation of high crude oil prices.

In France, a few years ago, President Mitterand almost apologized for the decision to reduce the number of construction starts of nuclear power stations from three per year to two. It was an issue of jobs, national pride, and self-sufficiency.

In Great Britain today, politicians are justifying raising electricity prices to improve the economics of building new power stations that use coal, at present highly priced, to subsidize the coal mining industry. Also, Great Britain is about to privatize electricity. The prime considerations are evidently not about demand energy efficiency.

The United States faces many political challenges on the energy scene. It consumes for its transportation needs half of the total energy used by all OECD countries for less than one-third of the people and also consumes significantly more energy per unit of gross domestic product (GDP) than any other country in the world, except Canada (Figure 2). Other than in the less developed countries, this ratio has fallen consistently since the early 1970s. The challenge now is to ensure that it continues to fall in the

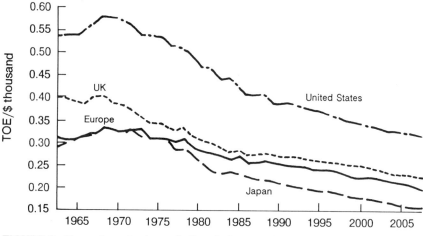

FIGURE 2 Energy intensity (tons of oil equivalent per thousand dollars).

future. Looked at another way, we need to extract more value, in terms of economic growth, from each unit of energy we consume; let us turn the index up the other way—more or less (Figure 3). There is much to do in the way of formulating policies that rekindle the public's incentives to use energy more efficiently. Then, to reduce increasing U.S. dependence on crude oil imports, we need to stimulate more indigenous exploration and development of known reserves.

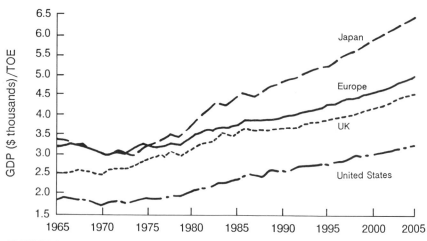

FIGURE 3 Productivity intensity (thousands of GDP dollars per ton of oil equivalent), the reciprocal of energy intensity (compare with Figure 2). High values of this indicator result when an economy is producing more with less energy.

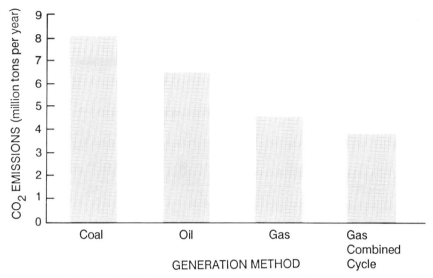

FIGURE 4 Carbon dioxide (CO_2) emissions for various fuels (million tons per year). Gas-fired combined-cycle generation facilities emit half as much CO_2 per gigawatt as conventional coal-fired facilities.

Environmental fears and the concern for the world ecosystem are global forces that can be harnessed to encourage energy efficiency. The most effective way of reducing atmospheric pollution both in power generation and in transportation—the main culprits—is to become more efficient at both. The less consumed the less emitted. This is a simple fact, yet public resolve has been allowed to weaken.

For electricity generation, ever-greater efficiency and cleaner fuels must be the objective. Methane is by far the best fossil fuel in this respect. It emits less carbon dioxide per unit of energy than any other fuel and generally produces no oxides of sulfur (Figures 4 and 5). The only count on which gas may perform less well than other fuels is in NO_x emissions, although these are, at worst, comparable with those of other fuels (Figure 6). Gas lends itself more readily to combined cycle generation, thereby raising the efficiency of generation from just under 40 percent to near 50 percent. Gas wins twice, resulting in about half the carbon dioxide emitted per kilowatt than coal. Not much is heard about this in Europe where gas may be underutilized for power generation.

Finally, in a brief review of global forces, there are the realities of microeconomics: that is, the criteria by which investment decisions are judged. Two effects act against greater energy efficiency:

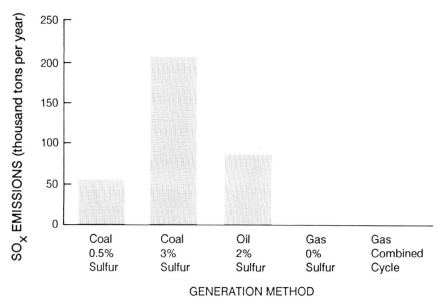

FIGURE 5 Sulfur oxide (SO_x) emissions, by fuel, for the configurations shown in Figure 4 (thousand tons per year).

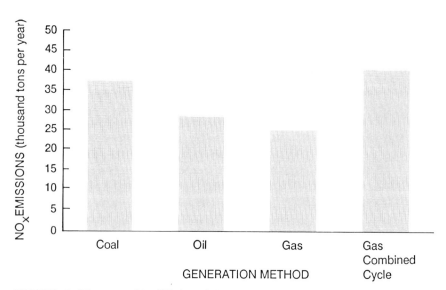

FIGURE 6 Nitrogen oxide (NO_x) emissions, by fuel, for the configurations shown in Figure 4 (thousand tons per year).

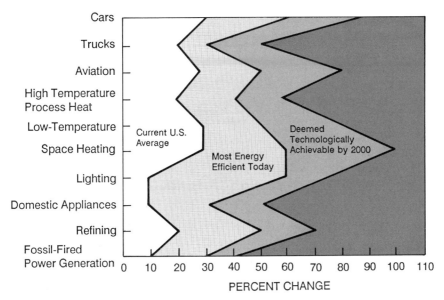

FIGURE 7 Current and potential improvement in end-use efficiency (percent change from 1973).

1. Investments in electricity generation, for example, are based on long lead times, utility rates of return, and payback periods of 15 to 20 years or more. On the other hand, decisions affecting energy demand, taken daily by millions of individuals and corporations worldwide, are based on very short payback periods of 3 to 6 years.

2. There is no way, at present, to reflect in these decisions their long-term consequences for both future supplies and the ecosystem. How can they be brought home to the public, to a present-day value of some sort, even if only qualitative, but nevertheless vivid and real?

Do not get the impression that nothing is happening with respect to greater efficiency. New aircraft are typically 20 percent more efficient than the stock average. The concept of the energy-efficient house is gaining ground, albeit slowly. In Europe the high-speed 185-mph train is developing and gaining popular appeal. It is more efficient, comfortable, and trouble-free than short air flights. The channel tunnel between England and France will be a further boost. But far more could be done, given the proper incentives, by harnessing existing technology as we develop future technologies (Figure 7).

If the current pace of demand continues and the current rate of improvement in energy efficiency is assured by the year 2020, more than twice as much total energy will be required as is used today. The bulk of

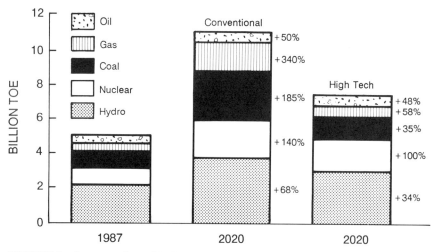

FIGURE 8 Current and possible future energy requirements for the noncommunist world (billion tons of oil equivalent). Without the potential benefits of technology, demand will be more than twice current levels by the year 2020.

this, it is forecast, can be met only by tripling the consumption of coal—the least elegant source of energy used widely today—and even this assumes that the use of nuclear power will more than triple. Yet, by harnessing the obvious benefits of technology, the outlook for world energy demand in 2020 could be radically different (Figure 8).

How can we reach the other, much more acceptable scenario—of achieving the same world economic growth over the next 30 years for not much more than current total energy consumption? This will only occur if greater prominence is given to energy demand issues and policies, and if engineers provide the lead.

The proposition of using less to produce more, both to conserve resources and to reduce waste, is at the very heart of all engineering philosophy. It is an objective that can find universal support. Greater efficiency, which facilitates greater growth, is a "virtuous circle" worth striving for and surely on the side of the angels.

I call on engineers because it is engineers who harness the extraordinary advances in science. "Science," said Von Karman, "discovers what is; engineers turn this knowledge into things that have never been." So engineers and technologists in general are best equipped to know what *can be* by using today's science and technology, and what *might be* by using tomorrow's.

Engineers who complain about the short-term attitudes of the public, financiers, accountants, and politicians have somehow allowed themselves to be painted into a corner, to become a service. They should be out there, in front, illustrating the opportunities and their benefits, both qualitative and quantitative. Consider how biotechnologist entrepreneurs have shown that hard-headed investors will put their money into "expectation." The price/earnings ratios of biotechnology stocks are about the future not about quarter-by-quarter results.

The challenge facing demand-driven energy policy options is how to influence short-term decisions to take into account long-term opportunity and potential penalties. In this we must seek help from economists. The challenge is similar to that which engineers and technologists constantly face in industry. So let me share with you a simple, powerful, equation imparted several years ago by a colleague; I use it in the task of harnessing technology for profitable growth.

"Change," said my friend, "is a function of dissatisfaction, vision, and a practical first step." *Dissatisfaction* involves the feeling that we can do better, rather than just complaining about how awful things are. *Vision* is, of course, what engineers and technologists can provide by articulating what might be. The *practical first step* is what engineers must fashion.

This is true also with energy. There is plenty of dissatisfaction and fear. The vision needs to be articulated and propagated by engineers, strongly supported by economists. What specifically should engineers do as a profession? The following are some suggestions.

First of all, the drive for greater energy efficiency should be at the top of all our agendas, and should remain there for the next decade. This is not the case today. Perhaps technical meetings could be used more effectively toward this end. Led by the National Academy of Engineering, the Council of Academies of Engineering and Technological Sciences brings together the academies of six nations, and several others are now eager to join this group. This mechanism could be used to have a strong voice, which would be heard worldwide.

Second, the excellent studies on energy conservation being carried out by many organizations should be actively supported: in the United States for example, this includes Harvard, Princeton, Berkeley, and the World Resources Institute. Perhaps a new ratio should be developed to measure the productivity of energy, such as gross national product (GNP) per unit of energy. This would heighten public awareness and thus could become a powerful force toward greater energy efficiency.

Third is for engineers and economists to devise means and measures to bring home to the public both the quantitative and the qualitative, long-term consequences, penalties, and benefits of day-to-day energy decisions. This is not an easy task, but it must be made to capture the imagination and support of the young, for it is their future we are talking about.

Perhaps a concept introduced by Professor Henry Jacobi of the Massachusetts Institute of Technology might help. He spent some time in Britain developing economic evaluation techniques for projects with long lead times, such as oil exploration. The concept is one of "options for the future," that is, to give a present-day value to the options for the future created by a decision made today—options which would not exist but for that decision. It is particularly useful for investments in new areas of business, new products, and new processes.

Other action might be taken to join with, or initiate, a worldwide review of public policy measures that have been successful in promoting greater efficiency through, for example, incentives, penalties, subsidies, and taxes. This would also deter policymakers, who might otherwise be tempted to use them, from those measures that have failed.

For example, the CAFE (corporate average fuel economy) legislation in the United States has been remarkably successful in raising the efficiency of U.S. automobiles. This was the only such policy in the world—and a very effective one—but the public seems to have turned its back on this. On the other hand, subsidizing energy to help the poor in less developed countries has not worked. Over the longer term, it has failed to alleviate poverty and has been a disincentive to energy efficiency.

Finally, even higher priority should be given to improving the safety of operation and the waste disposal of nuclear power stations. Nuclear energy is the cleanest of all fuels and produces no atmospheric waste. It is the ultimate answer to fears of the greenhouse effect, acid rain, and other forms of pollution. One is surprised that environmentalists do not promote it, demanding that it be made safer than it already is.

Such actions as these, plus setting out to understand the ecosystem more fully, seem the minimum that engineers should be actively promoting. If the engineering profession can be persuaded to slip into higher gear for more concerted and international action toward greater energy efficiency, and to assume its natural role as an agent of change, perhaps some words of warning from a wise "business" philosopher are in order. He said:

> There is nothing more difficult to carry out, nor more doubtful of success, nor more dangerous to handle, than to initiate a new order of things. For the reformer has enemies in all who profit by the old order, and only lukewarm defenders in all those who would profit by the new order. This lukewarmness arises partly from fear of their adversaries, who have the law in their favor; and partly from the incredulity of mankind, who do not truly believe in anything new until they have had actual experience of it.

Who wrote that, you may wonder? Schumpeter? Keynes? Friedman? Drucker? It was written by Machiavelli in *The Prince* (chapter 6), published in 1517.

NOTE

This discussion tacitly assumes world GNP growth of 3 percent per year and world population growth of 2 percent per year from the present until 2020.

REFERENCE

Goldemberg, J., T. B. Johansson, A. K. N. Reddy, and R. H. Williams. 1987. Energy for a Sustainable World. Washington, D.C.: World Resources Institute.

Contributors

HENRIK AGER-HANSSEN is senior executive vice-president of the Norwegian state oil company, STATOIL. He is also chairman of the board of the Norwegian Institute for Energy Technology and vice-chairman of the board of Rogaland Petroleum Research Institute. He is a member of the Royal Swedish Engineering Academy, the Norwegian Academy of Science and Letters, and the Norwegian Technical Academy of Science, and a foreign associate of the National Academy of Engineering. He holds an M.S. degree in nuclear engineering.

RICHARD E. BALZHISER is president and chief executive officer of the Electric Power Research Institute (EPRI) in Palo Alto, California. He is a member of the advisory boards of the Institute for Energy Analysis, the University of Michigan College of Engineering National Advisory Committee, the Academy Industry Program of the National Academy of Sciences, National Academy of Engineering, and Institute of Medicine, and the U.S. Department of Energy's Innovative Control Technology Advisory Panel. Dr. Balzhiser received his B.S. and Ph.D. degrees in chemical engineering and his M.S. degree in nuclear engineering from the University of Michigan.

WALLACE B. BEHNKE is vice-chairman of the Commonwealth Edison Company. Mr. Behnke serves on the boards of Commonwealth Edison and several affiliated corporations, including the Illinois Institute of Technology (IIT) and the IIT Research Institute, Duff & Phelps Selected

Utilities, the Materials Properties Council, and the National Planning Association. He is a member of the National Academy of Engineering, the American Nuclear Society, is a past president and honorary member of the Western Society of Engineers, and serves on the University of Chicago Board of Governors for Argonne National Laboratory and the Visiting Committee of the Nuclear Engineering Department of the Massachusetts Institute of Technology. Mr. Behnke received B.S. and B.S.E.E. degrees from Northwestern University.

PETER D. BLAIR is the manager of the Energy and Materials Program at the Office of Technology Assessment of the U.S. Congress. He is the author of many scientific reports, papers, and books, including *Multiobjective Regional Energy Planning, Geothermal Energy: Investment Decisions and Commercial Development* (with T. Cassel and R. Edelstein), and *Input-Output Analysis: Foundations and Extensions* (with R. Miller). He has served as a technical adviser to the National Research Council Panel on Central Electric Power Production and was an assistant professor of regional science and public policy at the University of Pennsylvania. Dr. Blair holds a B.S. degree in electrical engineering from Swarthmore College and an M.S. degree in energy management, an M.S.E degree in systems engineering, and a Ph.D. degree in energy management and policy from the University of Pennsylvania.

JOHN F. BOOKOUT is president and chief executive officer of Shell Oil Company. He is past chairman of the American Petroleum Institute and of the National Petroleum Council and is current chairman of the board of advisers for the Texas A&M University Institute of Biosciences and Technology. He is a member of the National Fish and Wildlife Foundation, the policy committee of the Business Roundtable, and the Conference Board. Mr. Bookout graduated from the University of Texas with B.S. and M.A. degrees in geology. He also holds honorary degrees of doctor of science from Tulane University and doctor of laws from Centenary College.

JOHN H. GIBBONS is the director of the Office of Technology Assessment of the U.S. Congress. He is a former director of the Energy, Environment, and Resources Center at the University of Tennessee and the first director of energy conservation for the Federal Energy Administration. He was chairman of the Demand/Conservation Panel for the National Research Council Committee on Nuclear and Alternative Energy Systems and has served on the Energy Research Advisory Board of the U.S. Department of Energy. Dr. Gibbons received a B.S. degree in mathematics and chemistry from Randolph-Macon College and a Ph.D. degree in physics from Duke University.

THOMAS E. GRAEDEL is a distinguished member of the technical staff of AT&T Bell Laboratories. His research interests are in solar physics, atmospheric chemistry and physics, measurement and chemical modeling of the earth's atmosphere, and corrosion of materials by atmospheric species. Because of his expertise in atmosphere-metal interactions, Dr. Graedel served as a consultant to the Statue of Liberty Restoration Project. He is associate editor of *Reviews of Geophysics*, a member of the governing board of the American Institute of Physics, and a member of the National Research Council Board on Atmospheric Sciences and Climate. Dr. Graedel received a B.S. degree in chemical engineering from Washington State University, an M.S. degree in physics from Kent State University, and both M.S. and Ph.D. degrees in astronomy from the University of Michigan.

JOHN L. HELM is an assistant professor of applied physics at the Department of Applied Physics and Nuclear Engineering at Columbia University. From 1987 to 1989 he was a fellow of the National Academy of Engineering. His current research interests include radiation damage to fission and fusion reactor materials, radiation transport and shielding, and energy systems and technology. He has also conducted research in solar energy, digital image processing, and coal gasification systems; he holds a patent for a novel method for coal gasification. Dr. Helm received a B.S. degree in chemical engineering and both M.S. and Ph.D. degrees in nuclear engineering from Columbia University.

WILLIAM F. KIESCHNICK is involved as a board member in a variety of industries including energy, aerospace, financial services, and new technology ventures. He is a trustee of a number of scientific, arts, educational, and humanitarian institutions. He was with the Atlantic Richfield Company (ARCO) from 1974 to 1985, moving through various assignments from production researcher to president and chief executive officer, the position from which he retired in 1985. Mr. Kieschnick received a B.S. degree in chemical engineering from Rice University and professional certificates in meteorology from the University of California, Los Angeles, and oceanography from Scripps Institution of Oceanography.

ROBERT MALPAS is managing director of the British Petroleum Company and chairman of BP Chemicals and BP Ventures. He has also served as president of Halcon Chemicals and held a variety of engineering and management posts, including chief executive and director of Imperial Chemical Industries. He is currently serving a term as president of the Society of Chemical Industry. Mr. Malpas received a first class degree in mechanical engineering from Durham University.

WILLIAM T. McCORMICK, Jr., is chairman of the board and chief executive officer of CMS Energy Corporation and its principal subsidiary, Consumers Power Company. Additionally, he has held policy-level positions in the Office of Management and Budget and the Energy Policy Office at the White House and served as a vice-president of the American Gas Association. He is currently on the board of directors of the American Gas Association and is chairman of the Gas Supply Committee and the Edison Electric Institute. Dr. McCormick received a B.A. degree in engineering physics from Cornell University and a Ph.D. degree in nuclear engineering from the Massachusetts Institute of Technology.

WILLIAM D. RUCKELSHAUS is chairman and chief executive officer of Browning-Ferris Industries. Mr. Ruckelshaus served as the first administrator of the Environmental Protection Agency from December 1970 to March 1973 and returned as its fifth administrator from May 1983 to January 1985. He is president of the Environmental Advisory Board of Control Resource Industries, Inc. Included among his board of trustee memberships are The Conservation Foundation/The World Wildlife Fund, the Urban Institute, the National Wildflower Research Center, and the Scientist's Institute for Public Information. He is chairman of the board of advisers of Riedel Environmental Services and is also the U.S. representative to the United Nations World Commission on Environment and Development. Mr. Ruckelshaus received his B.A. degree from Princeton University and obtained his law degree from Harvard University.

PETER H. SAND is senior environmental affairs officer with the United Nations (UN) Economic Commission for Europe, Geneva. Dr. Sand formerly was assistant director general of the International Union for Conservation of Nature and Natural Resources and deputy director of the Environmental Management Service of the United Nations Environment Program. He has served as UN consultant on environmental legislation to the governments of Colombia, Indonesia, Lebanon, and Niger. He participated in the drafting and negotiation of the 1976 Barcelona Convention for the Protection of the Mediterranean Sea Against Pollution, the 1979 Bonn Convention for the Conservation of Migratory Species of Wild Animals, and the 1985 Vienna Convention for the Protection of the Ozone Layer.

STEPHEN H. SCHNEIDER is head of Interdisciplinary Climate Systems Section at the National Center for Atmospheric Research. His current research interests include climatic change; food and climate, among other issues in environmental science and public policy; and climatic modeling of paleoclimates and of human impacts on climate, for example, carbon dioxide or greenhouse effect; and environmental consequences of nuclear war. Dr. Schneider is the author or coauthor of many scientific papers, proceedings, and books, including *The Genesis Strategy: Climate and Global Survival*, (with L. Mesirow), *The Coevolution of Climate and Life* (with R. Londer), and *Global Warming: Are We Entering the Greenhouse Century*.

He is a member of the Defense Science Board Task Force on Atmospheric Obscuration and a fellow of the American Association for the Advancement of Science. Dr. Schneider received a Ph.D. degree in mechanical engineering and plasma physics from Columbia University.

THOMAS C. SHELLING is Lucius N. Littauer Professor of Political Economy in the John F. Kennedy School of Government, Harvard University. Dr. Schelling has written extensively about conflicts between individual and collective behavior. He served in the Economic Cooperation Administration in Europe from 1948 to 1950 and in the Executive Office of the President of the United States from 1951 to 1953, and has been a consultant to the departments of State and Defense, the Arms Control and Disarmament Agency, and the Central Intelligence Agency. He has twice served as chairman of the Research Advisory Board of the Committee for Economic Development and is a trustee of the Aerospace Corporation. Dr. Schelling is a member of the National Academy of Sciences and the Institute of Medicine and is a fellow of the American Association for the Advancement of Science. Dr. Schelling received a B.A. degree in economics from the University of California, Berkeley, and a Ph.D. degree in economics from Harvard University.

JOHN W. SHILLER is Emissions Planning Associate with the Automotive Emissions and Fuel Economy Office of the Environmental and Safety Engineering Staff of Ford Motor Company. His activities include the design and development of pollutant-specific air quality monitors as well as air quality modeling and the evaluation of the need for and performance of emission controls and more recently, the impact of alternative fuel strategies on atmospheric ozone concentrations and global warming. He is chairman of the Atmospheric Chemistry Panel of the Motor Vehicle Manufacturers Association and has represented the Organisation Internationale des Constructeurs d'Automobiles at international forums on environmental issues. He received a B.S. degree in physics and an M.B.A. degree from the University of Michigan and an M.S. degree in physics from Wayne State University.

CHAUNCEY STARR was founding president and vice-chairman of the Electric Power Research Institute (EPRI). He is a former dean of the School of Engineering and Applied Science at the University of California, Los Angeles, following a 20-year industrial career. Dr. Starr's achievements include pioneering contributions to nuclear propulsion for rockets and ramjets, miniaturization of nuclear reactors for space, and development of nuclear-fission-powered electricity plants. He is a member and past vice-president of the National Academy of Engineering and a founder and past president of the American Nuclear Society. Dr. Starr received an electrical engineering degree and a Ph.D. degree in physics from Rensselaer Polytechnic Institute.

ALVIN M. WEINBERG is a distinguished fellow of the Institute for Energy Analysis, Oak Ridge Associated Universities, Oak Ridge, Tennessee. Dr. Weinberg was director of the Oak Ridge National Laboratory from 1955 to 1973; director of the Office of Energy Research and Development, Federal Energy Office, in 1974; and director of the Institute for Energy Analysis from 1975 to 1985. He is the author of *The Physical Theory of Neutron Chain Reactors* (with Eugene P. Wigner) and of *Reflections on Big Science*. Dr. Weinberg is a member of the National Academy of Engineering and the National Academy of Sciences. He has received numerous awards for his contributions to the design, development, and safety of nuclear reactors and the formulation of science policy. Dr. Weinberg received his B.S., M.S., and Ph.D. degrees in physics at the University of Chicago.

Index

A

Accidents and catastrophes
  Chernobyl, 189–190, 195, 197, 206, 266
  flooding, 83–84, 223, 227
  nuclear accident insurance, 200
  risk assessment, 224
  Three Mile Island (TMI), 190, 192, 194, 195, 196, 197, 265
  see also Flooding, greenhouse effect; Safety
Acid rain, 64
  chemistry of, 87–88, 89, 91, 94
  Europe, 14, 98, 99, 100–101, 104, 176, 246–247, 250–252
  global, 211, 212, 214
  regional, transboundary, general, 106, 246, 250–252
Aerosols, 94
  CFCs, 80, 94, 101, 102, 104, 218
Ager-Hanssen, Henrik, 15, 173–183
Agriculture, 65, 67
  climate and, 77, 223, 227, 235
  fertilizer, 67, 106
  plant genetics, 235
Air conditioning, 65, 67–68
  efficiency, 37
Aircraft, 235
Air quality and pollution, 95–98, 207
  aerosols, 94
  automobiles and, 98–99, 111–140
  carbon monoxide, 109, 115, 120, 121, 123
  CFCs, 80, 94, 101, 102, 104, 218
  economic development and, 95, 97–98
  electrification and, 63–64, 68
  Europe, 14, 98–101, 102–105, 246–259
  fossil fuels, 36, 63–64, 102, 184, 207, 211, 227, 247
  historical perspectives, 100, 102, 104, 117–118, 214
  international issues, 246–260
  models, 100, 122, 126–135, 247
  natural gas, 171–172
  ozone, 117, 120, 121–123, 124
  prediction methodology, 86–108
  smog, 98, 99, 100, 104, 117–118, 122, 130
  standards, autos, 113, 120–122, 123, 134–135, 139
  standards, international, 253, 258, 259–260
  transboundary, 14, 211, 246–260
  transition metals, 94

285

urban areas, 95, 98–99, 109, 111–140, 184, 207
  see also Acid rain; Carbon dioxide; Emission controls; Greenhouse effect; Nitrogen oxides; Sulfur dioxide
Alaska, 148–149
Alcohol fuels, 15, 113–114, 122, 124–126, 129, 131–135, 137, 269
Algeria, 173, 174, 179
Alternative energy sources, 11, 23, 85, 231
  alcohol fuels, 15, 113–114, 122, 124–126, 129, 131–135, 137, 269
  automobiles, 67, 68–69, 112–114, 122–138, 161, 234
  carbon monoxide, 65–66
  cofiring plants, 170–171, 176
  electrochemical cells, 65–66
  hydrogen fuels, 14, 15, 17, 65–66, 233, 234–235
  industrial boiler fuel switchability, 161–162
  synfuel, 30, 32
  wood fuel, 207
Ammonia, 67
Antarctic, 78, 83, 84
Atmospheric gases and trace constituents, 87–91, 99, 102
  carbon monoxide, 109, 115, 120, 121, 123
  CFCs, 80, 94, 101, 102, 104, 218
  transition metals, 94
  see also Carbon dioxide; Nitrogen oxides; Ozone, depletion; Ozone, formation; Sulfur dioxide
Atmospheric sciences
  chemistry, 87–94
  clouds, 82–83, 102
  stratosphere, 94, *see also* Ozone, depletion
  synoptic, 86–95
  troposphere, 102, 121, 255
  see also Air quality and pollution; Climatology; Greenhouse effect
Atomic Energy Act, 192
Atomic Energy Commission, 21, 22
Automobiles
  air pollution and, 98–99, 111–140
  alternative fuels, 67, 68–69, 112–114, 122–138, 161, 234
  computer simulations, emissions, 126–135
  diesel fuel, 114–115
  efficiency, corporate average fuel economy standards, 11, 14, 32, 38, 48, 276
  electric, 67, 68–69
  emission controls, 117–119, 123, 136
  energy efficiency, 37–38, 41
  fuel costs and efficiency, 40
  nitrogen oxide emissions, 117, 118, 121, 123, 125, 126, 129, 130, 134
  ozone emissions, 117, 120, 121–123, 124, 125, 129–130, 134
  standards, air quality, 113, 120–122, 123, 134–135, 139
  see also Gasoline
Automotive Consumer Profile, 137

B

Balzhiser, Richard E., 5, 16–17, 184–201
Batteries, 65, 244
  automobiles, electric, 67, 68–69
Beaufort Sea, 149
Behnke, Wallace B., 5, 12, 238–245
Belgium, 187, 189
Biotechnology, 235, 275
Blair, Peter D., 5, 9, 11, 12, 35–51
Bookout, John F., 4, 145–164
Brazil, 114, 122, 269
Building energy efficiency, 26, 47

C

California, 117–118, 120, 121, 130, 131–136, 139, 185
Calvert Cliffs nuclear plant, 190–191
Canada, 29, 168, 235, 246, 247, 251, 255
Cancer, 199
Capital investment
  electrification, 54–55, 242, 273
  emission controls, 256
  energy efficiency and, 47, 51
  gas- *vs* coal-fueled electric generation, 171–172
  nuclear power, 192, 198, 200
  oil industry, 153, 154–155, 160
  see also Industrial plants and equipment
Carbon dioxide, 7, 106, 165, 213–236
  atmospheric transport, 93, 94
  auto emissions, 115, 117
  conservation and, 33
  electric power generation, 63–64, 271

greenhouse effect and, 76–84 (passim), 93–95, 102, 184, 207, 213, 214–216
Carbon monoxide
  as fuel, 65–66
  pollution, 109, 115, 120, 121, 123
Catalytic converters, 118, 123, 258
Ceramics, 65
Chemistry
  atmospheric, 87–94, 131; *see also* Acid rain: chemistry of; Air quality and pollution; Smog
Chernobyl, 189–190, 195, 197, 206, 266
China, 79–80
Chlorofluorocarbons (CFCs), 80, 94, 101, 102, 104, 218
Cities, *see* Urban areas
Clean Air Act, 113
Climatology, 216–218
  drought, 223
  electrical load patterns and, 65
  manipulation of, 82–83
  models, 9, 76–78, 82, 102, 218–222
  population patterns, 65, 67–68
  projections, 9, 75–78, 82, 218–224
  regional, 219
  see also Greenhouse effect
Clouds, 82–83, 102
Coal, 22, 64, 104, 165, 184, 207, 233, 246, 269
  cofiring with natural gas, 170–171, 176
  Europe, 176, 256
  fluidized bed combustion, 185
  gasification, 65, 185
  gasoline, conversion to, 269
  nuclear generation *vs*. costs, 188
  shortages, 10–11
  synfuel, 30, 32
Cofiring plants, 170–171, 176
Colorado, 123
Committee on Atmospheric Sciences, 81–83
Committee on Electricity in Economic Growth, 53, 55
Committee on Nuclear and Alternative Energy Systems, 23, 36–38, 41, 224
Commodity prices, 145–146
Communications, 68, 69
Computers
  information sciences, 68, 244
  supercomputers, 130–131
Computer simulations

automobile fuels, air quality, 126–135
  chemical kinetics, 130
  climatologic, 8, 9, 76–78, 102, 218–222
  electrical systems engineering, 244
  marine, 78
Conservation, 12, 25–26, 239
  automobiles, 37–38, 68–69
  federal government, 35
  greenhouse effect and, 81
  national policy, 22, 28, 33
  social factors, 27
  see also Efficiency
Consumers
  automobile fuels, 136–138
  oil/natural gas, retail, 160–161
Consumption
  automotive petroleum, 116
  developing *vs* developed countries, 266
  electricity and GNP, 54, 55–63, 206–207
  historical perspectives, 7
  oil and gas, 163, 166, 174–177
  pollution associated with, 255–256
  see also Supply/demand
Continental shelves, *see* Offshore energy sources
Convention on Long-Range Transboundary Air Pollution (LRTAP), 252, 256, 257–258, 259–260
Cool Water project, 185
Cooperative Program for Monitoring and Evaluation of Long-Range Transmission of Air Pollutants in Europe, 246–247, 254, 259
Corporate average fuel economy standards (CAFE), 11, 14, 32, 38, 48, 276
Council for Energy Awareness, 198
Council of Academies of Engineering and Technological Sciences, 275
Court decisions, *see* Litigation
Czechoslovakia, 187, 254

D

Darmstadter, J. L., 98–100, 102, 104
Demand, *see* Supply/demand
Demography, *see* Population factors
Denmark, 176
Department of Energy, 32, 194–195
Deregulation, *see* Regulation/deregulation
Developing countries, 31, 207, 276
  aid to, 77
  electrification, 60–62

energy consumption, 266
Diesel fuel, 114–115
Disasters, *see* Accidents and catastrophes; Flooding, greenhouse effect
District of Columbia, 118
Drought, 223

E

Economic factors, 2–3, 267
  air quality and development, 95, 97–98
  carbon dioxide pollution, 79, 115–116, 223–224
  electrification, 68, 239–240, 241–242
  futures trading, oil, 146–147
  greenhouse effect, 223–224
  international pollution policy, 258
  natural gas, 153, 155, 168–171, 173–183
  nuclear power, 186–188, 189, 190, 191–192, 195–197, 198, 199
  oil industry, 145–163
  privatization, 176, 269
  projections, 276
  see also Capital investment; Consumers; Consumption; Efficiency; Foreign trade; Gross National Product; Industrial capacity and utilization; Prices; Production and productivity; Supply/demand
Efficiency, 265–276
  automobile operating costs and, 40, 67, 68–69, 112, 115, 137; corporate average fuel economy standards, 11, 14, 32, 38, 48, 276
  building energy efficiency, 26, 47
  capital investment, 47, 51
  cofiring plants, 170–171
  electricity and, 36–38, 46–47, 49, 54, 64, 184, 239
  greenhouse effect and, 231–232
  history, 14–15, 35–43
  household appliances, 36–38, 46–47, 49
  natural gas end use, 169, 170–171, 176
  oil industry, 49, 50, 149–151, 231
  prices and, 14, 36–38, 50, 231
  projections, 35–36, 43–51, 273–274
  supply/demand, 50, 273–274
  technology, 35, 39, 43
  see also Conservation
Electric power, 52–70, 184, 238–247
  air quality and, 63–64, 68
  automobiles, 67, 68–69
  batteries, 65, 67, 68–69, 244
  capital investment, 54–55, 242, 273
  cofiring plants, 170–171, 176
  deregulation, 241–243
  developing countries, 60–62
  electrochemistry, 65–66
  environmental implications, 62–66, 68, 184, 273
  Europe, 57
  GNP and, 54, 55–63, 239; global, 59–63
  historical perspectives, 23–30, 53–63
  natural gas, 65–66, 170–171, 176, 271
  plant cooling systems, 65
  policy, national, 26, 70
  population factors, 59, 60, 61, 65, 67–68, 69
  prices, 242–243
  projections, 52, 59–60, 61, 63–70, 206–207, 238–239
  regulation, 28, 41, 50, 241–243
  reliability, 240–241, 243–244
  safety, 243
  social factors, 60, 67–69, 70
  Soviet Union, 58, 60
  supply/demand, 238–240
  technology, 54–55, 61, 66–69, 70, 239, 244–245
  transmission, 65, 240–241, 242–244
  see also Hydroelectricity; Nuclear power
Electric Power Research Institute, 55, 65, 66, 193, 194, 195
Emission controls
  alternatives, 82–83, 94
  automobiles, 117–119, 123, 136, 138–140
  catalytic converters, 118, 123, 258
  greenhouse gases, 79–80, 81
  investment, 256
  nitrogen oxides, cofiring plants, 170–171
  see also Standards: air quality, autos
Energy for a Sustainable World, 267
Energy reserves
  natural gas, 28–29, 62, 147–149, 166–168, 173, 174
  oil, 147–149, 155, 156, 173
Environmental Protection Agency (EPA), 205, 208, 227
  air quality standards, 121, 135
  ambient pollution data, 119-121, 122–123
  corporate average fuel economy standards, 11, 14, 32, 38, 48, 276

Environment and pollution, 2, 3, 5,
    75–140, 205–212
  electrification, 62–66, 68, 184, 273
  energy efficiency and, 41
  Europe, 14, 97, 98–101, 102–105,
    246–259
  fossil fuels, 36, 63–64, 102, 184, 207, 211
  global, 8, 63–66, 75–84, 101–109, 198,
    205–206, 210–211; *see also*
    Greenhouse effect; Ozone, depletion;
    Ozone, formation
  historical perspectives, general, 7–10,
    208–209
  hydroelectricity, 62–63, 64
  legislation, 113, 117, 184–185, 191
  metal corrosion, 101, 103, 104
  natural gas, 15, 171–172, 233
  nuclear power, 6, 16, 32–33, 62–63, 64,
    80–81, 189–197, 194, 199, 207, 233
  regional forces, 85–109, 246–260
  waste management, 64, 66, 194, 199, 231
  see also Air quality and pollution;
    Climatology; Water systems and
    pollution
Ethanol, 113, 114, 122, 123, 269
Europe
  acid rain, 14, 98, 99, 100–101, 104, 176,
    246–247, 250–252
  coal, 176, 256
  electricity generation, 57
  environmental issues, 14, 97, 98–101,
    102–105, 246–260
  natural gas, 173–183, 256
  nuclear power, 176, 186–188, 195, 269
  oil demand, 8
  railroad, high-speed, 273
  see also specific countries
European Communities, 176, 235, 250,
    254–255
Exploration for oil, prices and incentive
    for, 50, 149–153, 157
Exports and imports; *see* Foreign trade

F

Federal action, *see* Government role
Federal Energy Regulatory Commission,
    241
Fertilizer, 67, 106
Finland, 187
Flooding, greenhouse effect, 83–84, 223,
    227

Food sciences, 67
Ford Foundation, 23
Ford Motor Company, 124
Forecasts, *see* Projections
Foreign countries, *see* Developing
    countries; Europe; International
    perspectives; Middle East; *specific
    countries*
Foreign trade
  climatically dependent crops, 235
  natural gas trade by Western Europe,
    173–183 (passim)
  nuclear facilities, 198
  oil imports, 11, 45, 49, 50, 158–159, 198,
    269; *see also* Organization of
    Petroleum Exporting Countries
Formaldehyde, 129–130, 135
Fossil fuels, 8, 14
  electrification and, 62–64
  environmental issues, 36, 63–64, 102,
    184, 207, 211, 227, 247
  GNP and consumption, 24, 163
  greenhouse effect, 63, 75–84 (passim),
    184, 211, 266
  see also Coal
France, 27, 28, 186, 187, 246, 269, 273
Frisch, J. R., 59–62
Futures trading, oil prices, 146–147

G

Gas, *see* Natural gas
Gasoline, 112–114
  alternative fuels, 67, 68–69, 112–114,
    122–138, 161, 234
  coal/natural gas, conversion to, 269
  corporate average fuel economy
    standards, 11, 14, 32, 38, 48, 276
  economics, *vs* natural gas, 169
  retail sector, 160–161, 169–170
  unleaded, 118
Gas Research Institute, 28–29, 169
Gas turbines, 65, 171, 234
General Public Utilities, 190
Genetic engineering, 235
Geology
  glaciology, 78, 83–84
  oil/natural gas, 149, 156–157
Geopolitics, 8, 70
  air pollution, transboundary, 14, 98, 99,
    100–101, 104, 106, 176, 211, 246–260
  see also Europe

Germany, Federal Republic, 79, 235, 245, 247, 254
Gibbons, John H., 5, 9, 11, 12, 35–51
Glaciology, 78, 83–84
Global systems and effects
  acid rain, 211, 212, 214
  efficiency and, 266–267
  electrification, 59–66, 68
  environmental, 8, 63–66, 75–84, 101–109, 198, 205–206, 210–211
  ozone depletion, 80, 101
  ultraviolet absorption, 101
  see also Climatology; Greenhouse effect; Ozone, depletion
Government role, 2, 14–15, 32
  energy efficiency, 35
  litigation, 190–191, 246, 260
  oil/natural gas lease sales, 153
  state-level action, 117–118, 123, 136, 171–172
  see also Legislation; Policy issues; Regulation/deregulation; Standards
Graedel, Thomas E., 4–5, 8, 85–110
Graz Seminar on the Control of Sulfur and Nitrogen Oxides from Stationary Sources, 256
Greenhouse effect, 9–10, 75–84, 101–102
  carbon dioxide and, 76–84 (passim), 94, 95, 102, 184, 207, 213, 214–216
  fossil fuels, 63, 75–84 (passim), 184, 211, 266
  international agreements, 79–80, 83, 211, 235
  methane, 15, 231, 233
  models, 7–8
  nitrogen oxides, 218
  risk assessment, 224, 233
Gross National Product
  electrification and, 54, 55–63, 239; global, 59–63
  energy and, 23–26, 39, 41, 45, 54, 55, 275
  oil and gas consumption and, 163
Gulf of Mexico, 149, 159

H

Haagen-Smit, Arie J., 117
Hayek, Fredrick, 32
Health Effects Institute, 135
Heat, see Thermal conditions and effects
Heat pumps, 65, 68
Helm, John L., 1–17, 213–237

Historical perspectives, 2–3, 10–11, 21–33
  air quality, 100, 102, 104, 117–118, 214
  auto fuels, 112–115
  climate change, effects, 76, 217–218
  electrical power, 23–30, 53–63
  energy efficiency, 14–15, 35–43
  environmental issues, general, 7–10, 208–209
  GNP and energy, 23–26, 39, 41, 45, 54, 55, 275
  liquid fuels, 114–115
  nuclear power, 190–191, see also Chernobyl; Three Mile Island
  oil/natural gas industry, 145–160 (passim), 174–175
  ozone levels, Europe, 100
  prices, 14
  see also Projections
The Horseless Age, 112
Household appliances, 46–47, 49
  air conditioners, 37, 65, 67–68
  heat pumps, 65, 68
  refrigerators, 36–37, 38, 67, 94
Hugo, Victor, 86
Human factors, 193–194, 267
  see also Management; Social factors
Hungary, 187
Hurricanes, 82, 83
Hydrocarbons, 218
  auto emission, 115, 117, 118, 123–126 (passim), 129, 130, 131, 134, 135
  see also Natural gas; Petroleum and petroleum products
Hydroelectricity, 24, 62
  environmental factors, 62–63, 64
  hydrogen fuel production, 235
  Norway, 176
Hydrogen fuels, 14, 15, 17, 233, 234–235
  electrochemical cells, 65–66

I

Imports and exports, see Foreign trade
Incrementalism, 28–30, 32–33
Independent power producers, 241–242
Industrial capacity and utilization
  electric power, 240, 243
  nuclear, 186–188, 207
  oil refineries, 147–149, 159–160
Industrial plants and equipment
  boiler fuel switchability, 162, 170
  Calvert Cliffs nuclear plant, 190–191

Chernobyl nuclear plant, 189–190, 195, 197, 206, 266
electrical transmission, 65, 240–241, 242–244
electrification and, 53–55, 67
electrochemical, 65–66
energy efficiency, 47
gas turbines, 65, 171, 234
nuclear, 186–188, 190, 192, 194, 195–196, 198
oil/gas industry, 149–151, 159–161, 166–167, 168, 178–179
pipelines, 178–180
power/fuel plants, 29–30, 192
Three Mile Island (TMI) nuclear plant, 190, 192, 194, 195, 196, 197, 265
Information sciences, 68, 244
Infrastructure, *see* Industrial plants and equipment
Institute of Electrical and Electronics Engineers, 243
Institute of Nuclear Power Operations, 193, 194, 195
Insurance, nuclear accident victims, 200
International Atomic Energy Agency, 186–189
International perspectives, 30–31
 agreements, policy criteria, 257–260
 air pollution, transboundary, 14, 98, 99, 100–101, 104, 106, 176, 211, 246–260
 greenhouse effect agreements, 79–80, 83, 211, 235
 hydrogen fuel, Canada-Europe agreement R&D, 235
 nuclear power, 186–188, 195, 198
 oil, foreign dependence, 6, 8–9, 55
 Organization for Economic Cooperation and Development, 23, 60, 79, 80, 250, 267, 269
 ozone, agreements, 80, 101; Europe, 100, 101, 104
 volatile organic compounds, agreements, 255
 see also Developing countries; Europe; Foreign trade; Global systems and effects; Middle East; Organization of Petroleum Exporting Countries; *specific countries*
Investment, *see* Capital investment
Irrigation, 65
Israel, 8

J

Jacobi, Henry, 276
Japan, 8, 28, 30–31, 186, 195, 209
Joint Committee on Atomic Energy, 33

K

Kellogg, William, 231
Keynes, Maynard, 32
Kieschnick, William F., 1–17
Korea, Republic of, 187, 195, 209

L

Labor unions, coal strike, 11
Lasers, 66–67
Law, *see* Legislation; Litigation; Regulation/deregulation; Standards
Lead, 118, 120
Legislation
 Atomic Energy Act, 192
 Clean Air Act, 113
 environmental, general, 117, 208–209
 National Environmental Policy Act, 191
 nuclear waste, 194
 Power Plant and Industrial Fuel Use Act of 1978, 50
 Price-Anderson Act, 200
 Public Utilities Regulatory Policy Act of 1978, 28, 241
 see also Regulation/deregulation; Standards
Liquefied natural gas, 15, 123, 178, 180, 182
Liquefied petroleum gas, 122
Liquid fuels, 41, 234
 automobiles, 112–114, 122–136
 see also Alcohol fuels; Diesel fuel; Ethanol; Gasoline; Methanol
Litigation, 190–191, 246, 260
Lovelock, James, 86
Lovins, Amory, 23

M

Machiavelli, Niccolo, 276
Malone, Thomas F., 81–83
Malpas, Robert, 6, 8, 12, 16, 265–277
Management
 gasoline, retail, 160
 nuclear power, 193

Manufacturing, *see* Industrial plants and equipment
Marine environments
  Beaufort Sea, 149
  greenhouse effect, 77, 78, 83–84, 211, 227
  Gulf of Mexico, 149, 159
  North Sea, 179
  Norwegian Continental Shelf, 174, 176, 178–180
  offshore energy sources, 148, 156, 174, 176, 178–180
  Prudhoe Bay, 148, 156
Mathematical models
  climatologic, 82, 218–222
  global electrification and GNP, 61, 62
  oil/gas economics, 153–158
  see also Computer simulations
McCormick, William T., Jr., 12, 15, 165–172
Mead, Margaret, 231
Metals, corrosion, 101, 103, 104
Methane, 102, 231, 234
  as greenhouse gas, 15, 231, 233
  see also Natural gas
Methanol, 15, 113, 123, 124–126, 129–130, 131–135, 137
Methyl tertiary butyl ether, 123
Metropolitan areas, *see* Urban areas
Mexico, 246
Middle East, 8, 50
  see also Organization of Petroleum Exporting Countries
Models, 13–14, 100
  air pollution, 100, 122, 126–135, 247
  greenhouse effect, 7–8
  systems theory, oil industry, 161–163
  see also Computer simulations; Mathematical models
Montreal Protocol on Substances That Deplete the Ozone Layer, 80, 101
Motor vehicles, *see* Automobiles

N

National Academy of Engineering, 275
National Environmental Policy Act, 191
National Petroleum Council, 153, 157–158
National Research Council
  Committee on Atmospheric Sciences, 81–83
  Committee on Electricity in Economic Growth, 53, 55
  Committee on Nuclear and Alternative Energy Systems, 23, 36–38, 41, 224
  Energy Engineering Board, 53
Natural gas
  cofiring with coal, 170–171, 176
  compressed, 123, 124
  economics, 153, 155, 168–171, 173–183
  efficiency of end use, 169, 170–171, 176
  electric generation technology, 65–66, 170–171, 176, 271
  environmental factors, 15, 171–172, 233
  Europe, 173–183, 256
  gasoline, conversion to, 269
  gas turbines, 65, 171, 234
  liquefied, 15, 122, 123, 178, 180, 182
  nitrogen oxides, 170–171
  prices, 153, 166, 169–170, 184, 188
  production, 147–149, 153, 155, 167–169
  projected role, 28–29, 153, 165–172, 174–176
  regulation, 171–172
  reserves, 28–29, 62, 147–149, 166–168, 173, 174
  safety, 171–172
  Soviet Union, 173, 174, 176, 179
  sulfur dioxide, 171
  technological advances, 156–157
  transportation, 168, 178–182, 183
The Netherlands, 122, 173, 195
New York, 122
New Zealand, 269
Nitrogen oxides, 272
  auto emissions, 117, 118, 121, 123, 125, 126, 129, 130, 134
  greenhouse effect, 218
  natural gas and, 170–171
  regional environment, 99, 100, 102, 246, 249, 251, 255, 256, 259–260
Nixon administration, 22
North American Electric Reliability Council, 239, 243
North Sea, 179
Norway, 173–174, 176, 178–180, 247
Nuclear power, 16–17, 30, 62, 165
  accidents, 189–190, 192, 194, 195, 197, 206, 256, 265
  Calvert Cliffs nuclear plant, 190–191
  capacity, 186–188, 207

Chernobyl nuclear plant, 189–190, 195, 197, 206, 266
economics and policy development, 14, 21, 22, 23, 27, 184–200
environmental issues, 6, 16, 32–33, 62–63, 64, 80–81, 189–197, 199, 207, 233
Europe, 176, 186–188, 195, 269
exports of facilities, 198
history, 190–191
plant cooling systems, 65
regulation, 191, 192–195, 196–197
safety, 16, 32–33, 63, 81, 189–197, 199, 207, 276
social factors, 190, 192, 197–198, 199
Soviet Union, 62–63, 187, 189–190, 195, 197, 206, 266
technological innovation, 193, 194, 195–196, 197, 198, 245
Three Mile Island (TMI) nuclear plant, 190, 192, 194, 195, 196, 197, 265
waste management, 194, 199
Nuclear reactors
advanced light water reactors, 195–196
liquid metal reactors, 196
modular high-temperature gas-cooled reactors, 196
Nuclear Regulatory Commission, 192, 193–195
Nuclear Utility Management and Human Resources Council, 193

O

Oceans, *see* Marine environments
Offshore energy sources
Gulf of Mexico, 149, 159
North Sea, 179–180
Norwegian Continental Shelf, 174, 176, 178–180
Prudhoe Bay, 148, 156
Oil, *see* Petroleum and petroleum products
Organization for Economic Cooperation and Development, 23, 60, 79, 80, 250, 267, 269
Organization of Petroleum Exporting Countries, 6, 8–10, 11, 44, 188, 198
Our Common Future, 206
Ozone, depletion, 214
Europe, 100, 101, 104
Montreal Protocol, 80, 101
stratosphere, 94

Ozone, formation
automobiles and, 117, 120, 121–123, 124, 125, 129–130, 134
urban areas, 99, 120
volatile organic compounds, 255

P

Paley Commission, 22
Pennsylvania
Three Mile Island (TMI) nuclear plant, 190, 192, 194, 195, 196, 197, 265
urban air pollution, Philadelphia, 130
Pesticides, 66
Petroleum and petroleum products
automotive use, 116, *see also* Gasoline
economics, 145–163
exploration, 50, 149–153, 157
foreign trade, 11, 45, 49, 50, 158–159, 198, 269; *see also* Organization of Petroleum Exporting Countries
investments, 153, 154–155, 160
liquefied petroleum gas, 122
prices, 2, 4, 8–11, 50, 145–147, 148–153, 155, 156–158, 160, 184, 188, 231
production, 147–153, 155, 159–160
projections, 153–156, 157
reserves, 49, 50, 147–149, 155, 156, 173
retail trade, 160
revenues, 151–153, 157
strategic petroleum reserve, 49, 50
sulfur, refineries, 159
synfuel, 30, 32
systems theory of, 161–163
Photovoltaics, *see* Solar energy
Pipelines, 178–181, 235
Planning, 6–7, 13–14
incrementalism, 20–30, 32–33
national energy system, 1–17
see also Projections
Plasma chemistry and physics, 66–67
Policy issues, 1–17, 276
automobile emissions, 136
carbon dioxide pollution, 224, 227, 231–236
conservation, 22, 28, 33
electrification, national, 26, 70
energy demand and, 39
greenhouse effect, 224, 227, 231–236
incrementalism, 20–30, 32–33
international agreements, 257–260

natural gas, Europe, 176
nuclear power, 22, 184–200
Politics
  carbon dioxide pollution, 223–224
  greenhouse effect, 223, 234–235
  nuclear power, 190, 192, 197–198, 199
  see also Geopolitics
Pollution, *see* Environment and pollution
Population factors, 6, 21, 45–46
  air quality and, 98
  climate and, 65, 67–68, 76, 79–80
  electrification, 59, 60, 61, 65, 67–68, 69
  see also Urban areas
Poro v. Lorraine Basin Coalmines, 246
Potential Gas Committee, 166
Power Plant and Industrial Fuel Use Act of 1978, 50
Price-Anderson Act, 200
Prices, 2–3, 6
  efficiency and, 14, 36–38, 50, 231, 268
  elasticity, 3–4, 39
  electricity, 242–243
  energy demand, 39
  exploration incentive and, 50
  GNP and, 41
  hydrogen fuel, 234
  natural gas, 153, 166, 169–170, 184, 188
  oil, 2, 4, 8–11, 50, 145–147, 148–153, 155, 156–158, 160, 184, 188, 231
Privatization, 176, 269
Production and productivity
  agricultural, 77
  electricity and, 23–26, 53, 239
  natural gas, 147–149, 153, 155, 167–169
  oil and gas, 147–149, 153, 155, 270
  see also Industrial capacity and utilization; Supply/demand
Project Independence, 26
Projections, 2–3, 26, 28–30, 32
  air pollution, methodology, 86–108
  climate, 9, 75–78, 82, 218–224
  economic evaluations, 276
  electric power, 52, 59–60, 61, 63–70, 206–207, 238–239
  energy efficiency, 35–36, 43–51, 273–274
  environment and development, 205–207
  Japan growth, 30–31
  natural gas, 28–29, 153, 165–172, 174–176
  oil industry economics, 153–156, 157
  supply/demand, 8, 21–23, 30–31

see also Planning
Prudhoe Bay, 148, 156
Public health, *see* Safety
Public Utilities Regulatory Policy Act of 1978, 28, 241
Putnam, P., 21–22

Q

Qualified generating facilities, 242

R

Railroads, 10, 68, 233–234, 273
Ray, Dixie Lee, 22
Refractory materials, 65
Refrigerators, 36–37, 38, 67, 92
Regional approaches
  acid rain, 106, 246, 250–252
  air pollution, transboundary, 14, 99, 100–101, 104, 106, 176, 211, 246–260
  climatic response, 219
  electrification, 60, 65
  environmental, 85–109, 109, 246–260
  greenhouse gases, 81
  nitrogen oxides, 99, 100, 102, 246, 249, 251, 255, 256, 259–260
  sulfur dioxide, 99, 101, 106, 260
Regulation/deregulation, 2
  electric power, 241–243
  emission controls, 117–119
  natural gas, 50, 166, 171–172
  nuclear industry, 191, 192–195, 196–197
  public utilities, 28, 50
  see also Environmental Protection Agency; Standards
Research and development, *see* Technological innovation
Retail trade
  gasoline, 160–161
  natural gas vs gasoline, 169–170
Risk assessment
  carbon dioxide pollution, 224, 233
  energy generation, 224, 225–226, 228–230
Ruckelshaus, William D., 5, 8, 11–12, 205–212

S

Safety
  electric transmission, 243
  natural gas, 171–172

nuclear power, 16, 32–33, 63, 81, 189–197, 199, 207, 276
  see also Accidents and catastrophes
Sand, Peter H., 5, 246–264
Scandinavia, 176, 180–181
  see also Denmark; Finland; Norway; Sweden
Schelling, Thomas C., 4, 9, 75–84
Schneider, Stephen H., 213–237
Seismic imaging, 149
Shell Oil Company, 146–147
Shiller, John W., 5, 111–142
Ships and shipping, liquefied natural gas, 180, 182
Smog, 98, 99, 100, 104, 117–118, 122, 130
Social factors, 5–6, 9, 268–269, 275
  carbon dioxide pollution, 223–224
  climate change, 65, 67–68, 219, 223–224
  demand, 7–8, 27
  electrification, 60, 67–69, 70
  labor unions, 11
  natural gas, 171–172
  nuclear power, 190, 192, 197–198, 199
  see also Consumers; Politics; Population factors; Safety
Soil sciences, 106–107
Solar energy, 30, 64, 185, 245
South Africa, 269
South African Coal, Oil, and Gas Corporation, 30
Southern California Edison, 185
Soviet Union
  air pollution, transboundary, 247
  Chernobyl, 189–190, 195, 197, 206, 266
  electrification, 58, 60,
  greenhouse gas control, 79
  hydroelectric power, 62–63
  natural gas, 173, 174, 176, 179
  nuclear power, 62–63, 187, 195
Standards
  air pollution, international, 253, 258, 259–260
  air quality, autos, 113, 120–122, 123, 134–135, 139
  building energy efficiency, 26
  corporate average fuel economy standards, 11, 14, 32, 38, 48, 276
  nuclear power, 193, 195–196
Starr, Chauncey, 7, 12, 52–71, 238
Statistical programs and activities
  EPA air pollution data, 119–121, 122–123
  European emission data, 247
Statutes, *see* Legislation
Strategic petroleum reserve (SPR), 49, 50
Stratospheric ozone, 94
  see also Ozone, depletion
State-level action
  auto emissions, 117–118, 123, 136
  natural gas regulation, 171–172
  see also specific states
State of Georgia v. Tennessee Cooper Company and Ducktown Sulphur, Copper and Iron Company, Ltd., 260
Sulfur dioxide, 165, 189, 218, 272
  coal, 165
  desulfurization, oil refineries, 159
  Europe, transboundary pollution, 246, 247, 248, 254, 255–257, 259
  natural gas, 171
  regional pollution, 99, 101, 106, 260
  see also Acid rain
Supercomputers, 130–131
Supply/demand, 2–7, 11–12, 26–27
  efficiency and, 50, 273–274
  electricity, 238–240
  GNP and, 23–26, 39, 41, 45, 61
  investment and, 6
  natural gas, Europe, 173–183
  oil, Europe, 8
  projections, 8, 21–23, 30–31
  technology and, 11
  see also Conservation; Consumption; Efficiency; Energy reserves; Production and productivity
Sweden, 176, 254
Switzerland, 247
Synergism, 17
Synfuel, 30, 32
Systems theory
  oil industry, 161–163
  see also Global systems and effects; Planning

T

Taiwan, 187, 195, 209
Tankers, liquefied natural gas, 180, 182
Taxes
  greenhouse effect, 235
  oil industry windfalls, 151, 163
  prices and, 14

Technological innovation, 3, 7–8, 12–13, 17, 33, 267–269
  biotechnology and plant genetics, 235, 275
  catalytic converters, 118, 123, 258
  climate manipulation, 82–83
  cofiring generating plants, 170–171, 176
  efficiency, 35, 39, 43
  electric power, 54–55, 61, 66–69, 70, 239, 244–245
  fluidized bed combustion, 185
  gasoline, retail sector, 160
  gas turbines, 65, 171, 234
  greenhouse gases, 81
  information sciences, 68, 244
  natural gas, 156–157
  nuclear power, 193, 194, 195–196, 197, 198, 245
  oil and natural gas, 149, 156–157, 158, 168, 169, 170, 180, 182
  solar energy, 30, 64, 185, 245
  supply/demand and, 11
  tankers, 180, 182
  see also Alternative energy sources
Television, 69
Texas, 121–122
Thermal conditions and effects, water pollution, 65
  see also Greenhouse effect
Three Mile Island (TMI), 190, 192, 194, 195, 196, 197, 265
Trail Smelter, 246
Transition metals, 94
Transportation
  aircraft, 235
  greenhouse effect and infrastructure, 233–234
  hydrogen fuel for public transportation, 235
  natural gas, 169, 178–182, 183
  pipelines, 178–181, 235
  railroads, 10, 68, 233–234, 273
  tankers, liquefied natural gas, 180, 182
  see also Automobiles
Troll Field, 179
Tropospheric ozone, 102, 121, 255

U

United Kingdom, 176, 180, 254, 273
United Nations, 60, 251, 253, 258, 259
Urban areas, 68, 69, 118
  air pollution, 95, 98–99, 109, 111–140, 184, 207
  greenhouse effects, 76, 81, 233, 234
  transportation, 234–235
U.S.S.R., *see* Soviet Union

V

Volatile organic compounds, 255

W

War
  Arab-Israeli War of 1967, 8
  World War I, 10, 11
  World War II, 32
Waste management, 66
  coal ash, 64
  landfill methane releases, 231
  nuclear, 194, 199
Water systems and pollution
  electric plant cooling, 65
  greenhouse effect, 223
  see also Acid rain; Hydroelectricity; Marine environments
Weinberg, Alvin M., 3–4, 21–34
White, Robert M., 69
Wood fuel, 207
World Association of Nuclear Operators, 195
World Bank, 60
World Commission on Environment and Development, 205–207, 209, 211–212
World Energy Conference, 31, 59–62